"In Praise For"

"Dr. Dan Schrage is our most knowledgeable and articulate champion of Army Aviation. No one understands its purpose, people and politics better than Dan" — Dr. Tom Thompson, Chief Engineer, Aeromechanics Division, U.S. Army Systems Readiness Directorate

"Dan's operational, acquisition and academic prowess establishes him as the preeminent rotorcraft developer. An outstanding leader and technical mentor." —Dr. William D. Lewis Former Director for U.S. Army Aviation Development

"As a decorated combat Army Aviator, accomplished aeronautical engineer, advanced systems innovator and past Director of the Army Rotorcraft Research Center of Excellence, Dr. Schrage brings an expert perspective in this important publication." – George T Singley III, Former Deputy Assistant Secretary of the Army (Research & Technology)

"Dan Schrage was an excellent athlete, my teammate, and captain of the 1966-67 Army Basketball Team with Bob Knight as Coach. He was our leader and our hardest worker and competitor" –Mike Krzyzewski (Ret.) Duke University Basketball Coach.

"Dan Schrage provides first-hand experiences and valuable lessons learned from the Cold War in Europe and the Vietnam War." --Paul Fardink Vertical Flight Society History Committee [LTC (Ret.)]

"AS USUAL GUARDIAN WAS PERFECT IN ALL RESPECTS"

A Memoir - Book 1 of 3

**A Full Lifetime Career of Seeking Perfection
Driven by Family and Mentors - A Trilogy**

Daniel P. Schrage

Requests for permission to make copies of any part of the work should be mailed to: Permissions Department, Publish Authority, 300 Colonial Center Parkway, Suite 100, Roswell, GA 30076-4892.

As Usual, Guardian was Perfect in All Respects: A Memoir - Book 1 in the trilogy A Full Lifetime Career of Seeking Perfection Driven by Family and Mentors

ISBN 978-1-954000-37-7 (Paperback)
ISBN 978-1-954000-38-4 (eBook)

All events, locales, conversations, and observations in this book are from the author's memories of them and from his perspective.

Editor: Bob Laning

Cover design lead: Raeghan Rebstock

Interior design: Teresa Evans

Published 2022, by Publish Authority

300 Colonial Center Parkway, Suite 100

Roswell, GA 30076-4892 USA

PublishAuthority.com
Printed in the United States of America

First edition

To my family, friends, and mentors.

PREFACE TO THE TRILOGY

"As Usual Guardian was Perfect in All Respects" is the first book in the Trilogy: *A Full Lifetime Career of Seeking Perfection Driven by Family and Mentors*.

The Trilogy is a tale of three careers: of military achievement, engineering accomplishment, and academic leadership. It illustrates how my striving for perfection has been driven by family and mentors through athletic and career advancement. It also supports, follows, and documents the changes in U.S. warfare, technology development, and academic transition in a changing world.

The Trilogy addresses my three careers, which build on each other, and overlap. Following the introduction and background for each of these careers, the three books summarize and highlight experiences and describe stories of my unique "Why" experiences. They also express my frustrations throughout these careers while striving toward perfection.

The first book begins with changes in the U.S. military as it transitions from the Cold War in Europe to the strategic mobility and nation-building efforts of the Vietnam War. As my motivation, the titles and words from John Denver's Definitive All-Time Greatest Hits will serve in the following reflection and theme for A Full Lifetime Career and many of the chapter titles in Book 1.

"I've been lately thinking about my life's time. All the things I've done. And how it's been. And I can't help believing. In my own mind. I know I'm gonna hate to see it end." from Poems, Prayers and Promises, John Denver's Definitive All-Time Greatest Hits.

The Trilogy follows with Book 2, "Development of the Next Generation of Army Aviation Systems." It documents the growth of air mobility through technology development over the next decade. My time at the Army Aviation Systems Command (AVSCOM) leveraged my operational and leadership experiences in Field Artillery and Army Aviation. The book is based on my technical education and training at outstanding civil and military institutions. This experience served as an entry into my second career as an aerospace engineer, technology manager, and senior executive, first with (AVSCOM) and then with the Aviation Research and Development Command (AVRADCOM).

This, in turn, led to my third career both as an academic rotorcraft design professor and as Director of the Center of Excellence in Rotorcraft Technology (CERT) at the Georgia Institute of Technology (GIT). I was also a consultant to industry and government.

With my third career, Book 2 transitions into Book 3, "A Graduate Program in Aerospace Systems Design". This Book will build off of my 1999 American Helicopter Society (AHS) Nikolsky Lecture, *Technology for Rotorcraft Affordability through Integrated Product and Process Development (IPPD)*." In Book 3, these technological advancements are brought into the academic environment with innovative IPPD methods and tools for the education of both civilian and military students. These advancements are then transferred to government and industry. The content of Book 3 comes from my experiences in Books 1 and 2 and my thirty-five years as a professor at the Georgia Institute of Technology (GIT).

Book 3 provides my experiences and accomplishments in these thirty-five years as a professor at Georgia Institute of Technology (GIT) or Georgia Tech. The final chapter in Book 3 focuses on my retirement years and completes my "Why" desire to write these books. It assesses my attempt to set the record straight for the three questions I identified for the Trilogy, one for each book.

Based on my experiences and insights during my three careers, I will help set the record straight on these three unanswered questions.

In my first career:

Book 1: Why was winning the Vietnam War not Successful? Could it have been Successful?

In my second career:

Book 2: Why is it so difficult for the Army to develop and field new aircraft?

And finally, in my third career:

Book 3: Why is it so hard for Academia to develop and implement new curricula?

The cover for Book 1 is a plaque given to me by the Men of the 13th Combat Aviation Battalion (CAB) Guardian S-3 Shop (illustrated in the picture below) to "The Guardian S-3" upon departure from Vietnam in January 1971. On the back of the Plaque is a note from the Guardian S-3 Staff. It reads,

"From The Men of the Guardian 3 Shop to The Guardian 3",

"As Usual Guardian was Perfect in All Respects."

This last statement is a relevant quote, typical of the 164th Aviation Group G-3 Shop following a successful 13th CAB operation in the Mekong Delta in 1970. While MAJ Burnam Melton served figuratively as the S-3, I, CPT Daniel P. Schrage, ran the S-3 Shop as a junior Captain and officially became the S-3 when the 13th CAB moved to Can Tho in November 1970. This followed the transfer of the Soc Trang Army Airfield to the Vietnamese Air Force (VNAF). MAJ Melton was then assigned by the 13th CAB Commander as the 221st Recon Aircraft Company Commander. He was an excellent officer and was very supportive of me as the 13th CAB operational S-3, as he recognized he had little helicopter operational experience.

Men of the Guardian S-3 Shop, 13th CAB

At the Center, sitting down in the above picture, is MAJ Burnum Melton, S-3, with me on his left-- CPT Daniel P. Schrage, Asst S-3/ S-3. Comments from MAJ Melton and MAJ Charles A. Lepore in my final South Vietnam Officer Efficiency Report (OER), July 1970-January 1971, support my strive for perfection, as written by MAJ Burnum E. Melton, OER Rater, and MAJ Charles A. Lepore, 13th CAB Executive Officer, OER Endorser. They both state that I, CPT Daniel P. Schrage, was "the most outstanding officer they had known." For me, this recognition, along with my accomplishments, helped validate that the 13th CAB in 1970-71 strived for and accomplished perfection.

Book 1 is entitled, "As Usual, the Guardian was Perfect in All Respects." The words "Perfect in All Respects" have had a special meaning and have driven me throughout my lifetime. I have been driven to strive for perfection by a highly motivated family and mentors throughout my "Full Lifetime Career."

This provides the "Why?" for this book as well as for the other two books in this trilogy

Book 1 Outline

General Structure:

Each chapter starts with an introduction and a background summary of my experiences, followed by my unique stories. They include some events for the organizations and men I served.

Chapter One: Introduction

1. Provides rationale for the emblem and title for the book, along with a brief description of the follow-up chapters in Book 1
2. Provides an overview of the Trilogy of books to accomplish the "Why?"

Chapter Two: Unique Experiences in South Vietnam and Cambodia

1. Arrival In South Vietnam and Assignment to the Mekong Delta
2. Brief History of the 162nd AHC and my first assignment
3. Experiences and stories with the 162nd AHC from January 1970-July 1970
4. A brief history of 13th CAB and Significant Stories
5. Experiences as Asst S-3/S-3 in 13th CAB and My Stories, July -Nov 1970
6. Activities out of Can Tho AAF, R&R and Home, Nov 1970-Jan 1971

Chapter Three: The Author's Growing Up and Moving Forward

1. Growing up in the Midwest
2. Experiences as a pseudo "Tom Sawyer" or "Huck Finn"
3. Experiences as a rising athlete in baseball and basketball
4. High School experiences at Mater Dei High School, Breese, IL.
5. College Experiences at Quincy College, Quincy, IL
6. Experiences as a cadet and athlete at USMA, West Point, NY

Chapter Four: The Author's Experiences in Schools and as Nuclear Weapon Battery Commander

1. Experiences in military courses from field artillery basic course, ranger school, and nuclear warhead school
2. Experience as a nuclear weapon battery commander in West Germany, 1968-1969
3. Experiences from Rotary Wing Flight School and Deployment to South Vietnam
4. References

Chapter One

INTRODUCTION TO BOOK 1

The first book tells a story about some truly unique and rewarding experiences I encountered growing up, during the eras of the Cold War and the Vietnam War, and over three careers. I saw tremendous changes in American culture, military strategy, and rapid technological advancement during this time.

This Shield of the Mekong (Figure 1) was the emblem worn proudly by members of the 13th CAB in South Vietnam. They were the "*Guardian of the Mekong.*" Their mission in 1970-71 was to defend and support the Army of the Republic of Vietnam (ARVN), Special Forces, the Navy Seals, and other indigenous South Vietnam ground forces in their war against the Viet Cong (VC) and the North Vietnamese Army (NVA) or People's Army Viet Nam (PAVN).

The major U.S. ground force in the Mekong Delta, the 9th Infantry Division, left in 1968-69, leaving the 164th Combat Aviation Group (CAG) and its CABs as the primary U.S. defenders in the Mekong Delta. The 13th CAB, as part of the 164th Combat Aviation Group (CAG), consisted of four assault helicopter companies in 1970, a fixed-wing recon airplane company, and an air cavalry troop, also identified in Figure 1.

Guardians of the Mekong

162nd Assault Helicopter Co.
121st Assault Helicopter Co.
191st Assault Helicopter Co.
221st Recon Airplane Co.
336th Assault Helicopter Co.
C Troop, 16th Air Cavalry

Figure 1. 13th CAB Units, 1970-71

The 13th CAB was designated as the Task Force Guardian by the 164th CAG with direct support to the 21st ARVN Division and an area of operations (AO) south and west of the Mekong River, as shown in Figure 2. Two other CABs, the 214th, and the 307th, also had AOs north and east of the Mekong River. All three CABS provided aviation support in the IV Corps Tactical Zone (IVCTZ) up to Saigon.

Figure 2. General Map of Mehong Delta

The book title is "As Usual, the Guardian was Perfect in All Respects," which is about how the 164th Combat Aviation Group (CAG) G-3 Staff often expressed their thoughts after a successful 13th CAB mission. The words Perfect in All Respects" have also had a special meaning for me throughout my lifetime.

"Seeking Perfection" is the "Why?" for this book, as I have been encouraged to strive for perfection by highly supported family and mentors throughout my three careers. They also strived for perfection in their different areas. Hence, I have sought similar perfection.

An example starts with my mother, Mary, striving for home cleanliness perfection as a homemaker while still working as an excellent full-time teacher. My father, Albin, taught me in the seventh and eighth-grade classes and was also my basketball and baseball coach. He taught me to excel in class and to seek perfection in sports.

Striving for perfection in sports at an early age resulted in me playing for and serving as the West Point USMA basketball team captain in 1967 for a world-renowned, notorious college basketball coach, Bob Knight. He strived for perfection on the court and motivated his teams to achieve it.

A third example was a senior government executive, Charles C. Crawford, who served as my engineering mentor for next-generation Army Aviation Aircraft development. He has been called the "Father of the Army's UH-60 Black Hawk Helicopter" by both government and industry. Crawford strived to make Army Aviation development successful by insisting every presentation chart prepared was "perfect in all respects" before delivering it to potential decision-makers. It often included weekends fine-tuning these charts. His striving for perfection struck a chord with me throughout my career. I especially related to it before and after my time in South Vietnam, as experiences in Vietnam motivated and re-enforced this striving for perfection.

Chapter two of this book covers my unique, challenging, and rewarding experiences in South Vietnam and Cambodia as an Army Aviator-first in the 162nd Assault Helicopter Company (AHC), and then with the 13th Combat Aviation Battalion (CAB).

In chapter three, I reflect on my experiences growing up in the Midwest, moving forward to the United States Military Academy (USMA), and accepting a commission as a second lieutenant in Field Artillery. After attending the Field Artillery Basic Course and Ranger School, I was assigned as an active duty military officer fighting the Cold War in Europe. It included being commander of a nuclear missile battery, 1st Battalion 34th Artillery, in Munich and Augsburg, West Germany. This tour was during the Warsaw Pact (Soviet Union) 1968 invasion of Czechoslovakia. I also participated as a battery commander in the first Return of Forces to Germany (REFORGER I) for the 24th Infantry Division Exercise in 1968-69. Chapter three ends with my returning to the USA for Army Aviation Flight School and en route to South Vietnam which leads back to Chapter 2.

Chapter four begins with my return from South Vietnam and reuniting with my growing family. After my return, I attended the Field Artillery Officer Advanced Course at Fort Sill, OK, and was selected for Advanced Civil Schooling in Aerospace Engineering at Georgia Tech.

In 1974, I graduated with a Master of Science (MS) degree in Aerospace Engineering. My next assignment was to the Army Aviation Systems Command (AVSCOM) in St. Louis, MO.

Unfortunately, my father passed away from a brain tumor at age 61 in our Southern Illinois home while I was enroute to this new assignment to AVSCOM. Chapter 4 leads to and transitions to Book two, *Development of Next Generation of Army Aviation Systems.* At AVSCOM, I served as an aerospace engineer, technology manager, and senior executive in developing the next generation of Army Aviation Systems.

Chapter Two

UNIQUE EXPERIENCES IN SOUTH VIETNAM-CAMBODIA

"Leaving on a Jet Plane- Don't Know When
I will be back Again" — John Denver

When we left for the Vietnam War, most of us didn't know when or if we would be back again.

The experiences began when I arrived in South Vietnam (VN) in January 1970. I served as a combat helicopter pilot, a lift ship platoon leader, an operational Air Mission Commander, and briefly, as the gunship platoon leader in the 162nd Assault Helicopter Company (AHS) in Can Tho, VN. After six months, I was chosen as the Assistant S-3 and then as S-3 Operations Officer for the 13th Combat Aviation Battalion (CAB) in Soc Trang, VN.

The Soc Trang Army Airfield was transferred to the Vietnam Air Force (VNAF), the first transfer under the Helicopter Vietnamization Program, on November 4, 1970. The 13th CAB was relocated back to Can Tho, VN, its original home.

Throughout all three books, the helicopter has played a significant role in my life and careers. What I learned from flying helicopters and developing and running airmobile operations in VN and Cambodia, set the stage for my follow-up careers. These careers were focused on further helicopter development and modernization, including their operational usages, through teaching and research. The helicopter was the foundation and thread throughout all my careers in many ways. From my experiences in Field Artillery and REFORGER I in Germany, I learned that the third dimension, "air," was what the helicopter could bring to and revolutionize the battlefield.

Arrival In South Vietnam and Assignment to the Mekong Delta

I arrived in Vietnam at Tan Son Nhut Air Base in Saigon following a chartered commercial flight from Oakland, CA, in January 1970, with other Army aviators and soldiers. Three of us were transferred to Can Tho, South Vietnam, by a C-123 aircraft. We were put up in barracks for the next few days before being assigned to specific units. Much time was spent in the Can Tho Army Airfield Officer's Club, meeting other Army aviators and discussing their experiences. We were thankful that we went south to the critical Mekong Delta Region. Army Aviation and its Air Mobility were expanding, and we could be part of it. Brigadier General George W. Putnam, Jr., who took command of the 1st Aviation Brigade on 6 January 1970 from Major General Allen M. Burdett, Jr., remarked,

The real story of the Aviation Brigade is in the 164th Group in the Delta. Elsewhere it was, 'give so many helicopters here; and so many there.' The CG, 1st Aviation Brigade, exercised very little control over the assets of the Brigade in the I, II, and III Corps. But the 164th Group was not precisely controlled. Its commander could move assets: organize task forces, etc. They had a combat organization that permitted a diversity of aviation assets to support three ARVN Divisions, the 7th, 9th and 21st" and the Regional and Popular Forces, called "Ruff/Puffs."

In December 1969, Colonel William J. Maddox, Jr., was assigned as Commanding Officer of the 164th Aviation Group after commanding the 3d Brigade of the U.S. 25th Infantry Division. It was a good choice since Colonel Maddox had extensive experience in the Delta. First, he was Commanding Officer of the 13th CAB from July 1965 to August 1966. Second, he was Senior Advisor to the 21st Army of the Republic of Vietnam (ARVN) Infantry Division in the Mekong Delta from September 1966 to June 1967.

Besides the unique geographic features, the big difference between the operations in the IV Corps Tactical Zone and the other areas of Vietnam was the lack of any long-term division-size U.S. troop commitment. This made the aviation group commander in the Delta, in a critical sense, the "airmobile commander." (Vietnam Studies: Air mobility, 1961-1971, Dept of Army). The Mekong Delta was the breadbasket for South Vietnam (about 75% of rice produced) and had about 60% of the South Vietnam population. There was a strong feeling that following the 1968 Tet Offensive until 1972; the Vietnam War was being won, especially in the Mekong Delta and a few other areas around Saigon..

Previously, the approach by General Westmoreland was to build up more and more the commitment of U.S. Forces, asking annually for more and more U.S. troops, often at 200,000 levels, in the years 1965-68. He also pushed for extensive combat *search and destroy* operations, mainly with U.S. Forces. He often did not effectively incorporate the ARVN divisions or the use of Ruffs/Puffs, as the objective was not to hold territory or secure populations. The victory was assessed by a higher enemy body count.

In the wake of Tet 1968, new tasks were confronting the new leadership triumvirate of General Abrams, Ambassador Ellsworth Bunker, and William E. Colby (a career officer of the Central Intelligence Agency (CIA) who had earlier been the Agency's Chief of Station, Saigon). Their new political-military strategy was to be *clear, hold* and *build*: to clear areas from insurgent control, to hold them securely, and to build durable, national Vietnam institutions. This approach accommodated nation-building, something not necessarily espoused by GEN Westmoreland. (From Book, "A Better War: The Unexamined Victories and Final Tragedy of America's Last Years in Vietnam", by Lewis Sorley).

In response to the "A Better War" approach in Sorley's book, General H. Norman Schwarzkopf, U.S. Army Commander of the Iraqi War (and an instructor of the author at USMA), stated:

"An extraordinary piece of work that is bound to become a valuable part of historical documentation about the war in Vietnam. It answers many questions that have been too long on the minds of all of us who fought in that war. It also is first to **set the record straight** concerning the outcome of that conflict".

Another more current relevant reference, *Foreign Relations, 1969–1976, Volume VII. Vietnam July 1970-January 1972*. Special National Intelligence Estimate1SNIE 57-70, was released in SEPTEMBER 2010. It illustrated that there was an opportunity to win the Vietnam War. On 12 March 1970, while Prince Sihanouk was out of the country General Lon Nol of Cambodia sent an official message to the North Vietnamese demanding that their forces be withdrawn from the country within three days. Six days later the Cambodian National Assembly passed a resolution deposing Sihanouk. Lon Nol as Prime Minister took charge of Cambodia's affairs, and the stage was set for yet more drastic changes to the war in Southeast Asia. This opportunity could have been exploited with a coalition of Cambodian, South Vietnamese and U.S. Forces. It was a missed opportunity for "Winning the Vietnam War" through a Better War coupled with a more defendable Region in Southern Cambodia including Routes in and out of Phnom Penh and in the Mekong Delta including Saigon and Binh Hoa, essentially the III and IV CTZs. The essentials of how this could be achieved are included in the Foreign Relations Reference above and should be explored in another book.

The following discussion will familiarize the reader with the Mekong Delta, its hot spots and the 162nd Assault Helicopter Company (AHC) operations which I participated in from February to July 1970. This included helping plan and participate with the 164th CAG airmobile combat assault of the 9th ARVN Division in the Cambodia incursion.

A map of the Mekong Delta, 1970-71, with key participants, landscapes and locations is shown in Figure 3 and will serve as a roadmap for Chapter two.Many of these details are discussed in the stories, providing a holistic view of Army Aviation in the Mekong Delta, 1970-71. The terrain was primarily covered with rice paddies which produced nearly three-fourths of all rice grown in South Vietnam, similar to almost everywhere you traveled. Canals and tree lines crisscrossed rice paddies. However, as highlighted as * * * in the Mekong Delta Map, Figure 3, there were critical areas from which the Viet Cong and North Vietnam Army (NVA or PAVN) infiltrated and operated. The Seven Sisters Mountains along the Vietnam and Cambodia Border was a key hotbed, illustrated in Figure 4.

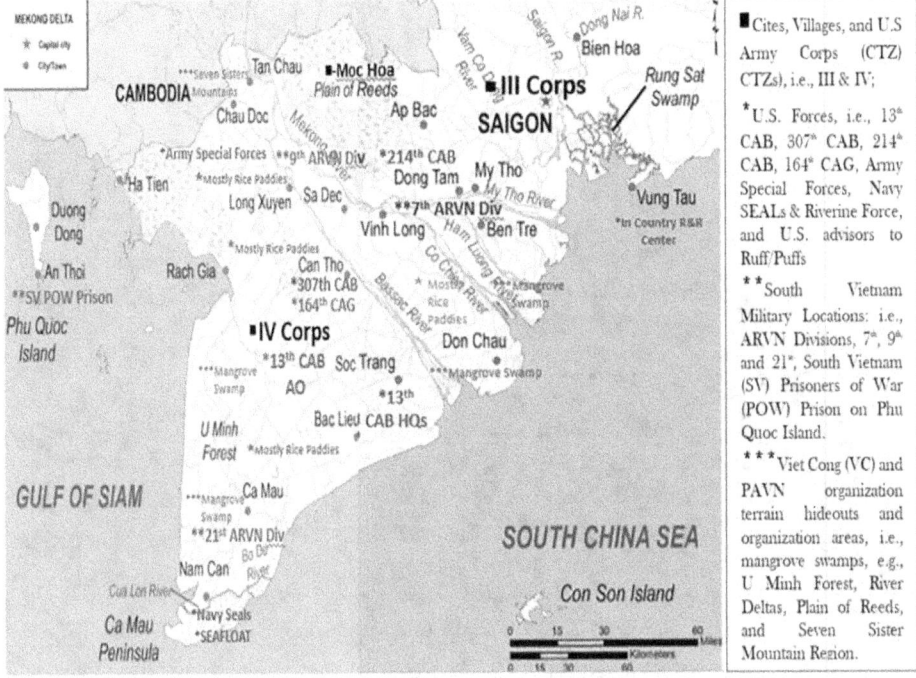

Figure 3 Roadmap for Key Participants, Landscapes, and Locations in IV Corps Tactical Zone (CTZ)

Figure 4 Seven Sisters Mountain Region

Figure 5 U Minh Forest Mangrove Swamp

These were staging points for PAVN units' entry into the Mekong Delta, often crossing over to the U Minh Forest, their primary staging and support area. During the Vietnam War, U.S. Special Forces Camps were set up on several of these mountains for surveillance and tracking. The 164th Combat Aviation Group (CAG) provided helicopter support to these camps. It was often tracked by PAVN radar-controlled anti-aircraft weapons, requiring helicopter low-level maneuvering flight rather than the higher than 1500 feet flight level typically used to avoid small arms fire.

The Seven Sisters Mountains Region, shown in Figure 4, was a PAVN/Viet Cong stronghold for entry into South Vietnam from Cambodia. During 1970 NVA and PAVN units crossed the Mekong Delta to U Minh Forest, Figure 5.

Figure 6 Nui Coto Mountain

A STORY ON SPECIAL FORCES OPERATIONS (NUI COTO AND THE MIKE FORCE ASSAULTON TUK CHUP), AN EXAMPLE OF SPECIAL FORCES OPERATIONS

The southern-most of the Seven Sisters Mountains is a rugged peak called Nui Coto, Figure 6. Early in 1969, representatives from the 5th Special Forces and the 44th Special Tactical Zone decided to secure the region. The Viet Cong and PAVN strongholds atop Nui Coto, and Tuk Chup, a lesser rockpile peak to the southwest, needed to be attacked and destroyed to secure the Seven Sisters Mountains Region.

The Nui Coto fortress was reputed to be the headquarters of Chau Kim, a notorious, almost mythic Cambodian communist. Chau and his troops had terrorized local farmers for years, demanding food and supplies for their camps.

Phase I of the operation called for The Civilian Irregular Defense Group (CIDG) troops (indigenous mountain tribe members), soldiers from Special Forces A-Camps in the area, and National Police to cordon off and search local villages. That would cut off much of the enemy's food supply, liberate the local populace, and help identify any local VC. camps in the area.

Figure 7 - 5th Special Forces Enroute to Tuk Chup

Phase II called for 1st and 2nd Battalions of the 5th Mobile Strike Force Command(MSFC) (Nha Trang) to move in and seal off Tuk Chup (chup means knoll), as illustrated in Figure 7. The knoll, covered with huge rocks and boulders the size of two-story houses, was a promontory of the Nui Coto group and was riddled with caves used as storage facilities by the VC and PAVN. Those caves would be a source of resupply and reinforcements if not isolated from the prominent peak.

Finally, in Phase III, Tuk Chup would be cleared of enemy forces before friendly elements moved in to clear and secure Nui Coto itself. It was a simple plan, but there was nothing simple about Nui Coto or Tuk Chup. Phases I and II were completed by 14 March 1969, and the Mike Force units were flown into positions west of Nui Coto the following day. At 0430 hours on 16 March 1969, an artillery barrage commenced, followed by the ground assault an hour later. In the predawn haze, hundreds of Mike Force soldiers with their Green Beret advisors moved silently across the paddies and marshland surrounding the ominous peak (Figure 7). Sgt 1st Class Richmond Nail and the troops of 6th Company led the way, followed by Sergeant John Talley and his 4th Company. Hidden within the 800-foot-tall pile of boulders, enemy sharpshooters watched and waited. Special Forces SFC Albert Belisle, commanding one of the Strike Force companies of Montagnard, said, "When we crawled over the big boulders, Viet Cong sharpshooters picked us off one by one. We tried to crawl forward under the rocks, but the Viet Cong came on top of us, throwing and rolling grenades down." There were snipers everywhere. Belisle's Company had lost two men to sniper fire before

Figure 8 5th Special Forces Enroute to Tuk Chup

they reached the foot of the knoll. As the MSFC troops attempted to push beyond the massive rocks at the mountain base, they encountered even stiffer resistance. "...We fought straight up the hill," Belisle said. "[But] every 20 yards another man fell.", Figure 8. Artillery Fire on Tuk Chup, Figure 9.

Figure 9 Artillery Fire on Tuk Chup

As his unit struggled to move, SFC Belisle asked for three volunteers to crawl out and attempt to infiltrate the Viet Cong camp. His three Montagnard platoon leaders volunteered. Lieutenants Ha-Elyz, He-Non, and Mang-Son took off their web gear and helmets and handed their M16s to the Green Berets.

Then, sticking knives in their belts, they stuffed their pockets with grenades and slipped away between the rocks. Upon reaching what appeared to be a main entrance to the Tuk Chup cave complex, the three Montagnard officers killed the guard and stole inside. Once inside, they moved from room to room, surveying the accessible areas of the cave system. If they encountered VC, they bluffed their way past in their black-covered bodies and kept moving. Before departing the caves, they grabbed as many documents as they could safely carry and made their way back down the rocky bluff. With the sun slowly dying behind the peaks of southern Cambodia, the battle continued to roar, and the three youngsters crossed safely back into the area controlled by their Company some eleven hours after they had left. "They got us all the poop we needed to get the company into the caves," Belisle said proudly, "...I never thought I'd see the three alive again, but ...they were back, with eleven Charlies killed and [they] still [had] seven grenades left." As a result of the efforts by the three Montagnard platoon leaders, the Tuk Chup "rock pile" and subsequent Nui Coto were cleared of VC and PAVN, at least for the time being. (Battle for Nui Coto Mountain). This should have been followed up by the Better War functions of *clear, hold and secure* but wasn't.

The Mangrove Swamps, especially the U Minh Forest, called "The Forest of Darkness," is illustrated in Figure 5. The U Minh Forest was the primary support base for the Viet Cong and North Vietnam Army (PAVN) in the Mekong Delta, dating back to the First Indo-China War. It was the second-largest mangrove swamp in the world, after the Amazon River Basin Mangrove Swamp. It was also never completely controlled by the ARVN forces throughout the Vietnam War. In fact, in the First Indo-China War, a French paratroop battalion dropped into the U Minh Forest in 1952, and the ~500 paratroopers were never heard from again. They have been called the "Lost French Battalion." In the Fall of 1970 and Spring of 1971, the 21st ARVN Division conducted significant operations in the U Minh Forest, supported mainly by the 13th CAB. They found some remnants of the "Lost French Battalion," including a unique 75 mm canon. Also uncovered was a large VC/PAVN POW Camp built underground. There was evidence that U.S. prisoners were held there. When a senior advisor, a U.S. Army Colonel, to the 21st ARVN Division was killed by VC/PAVN, the 21st ARVN withdrew from the U Minh Forest and never returned. This is another example of where the Better War functions of *clear, hold and secure* should have been implemented.

A Story on American POWs Left Behind
(reported U.S. Prisooners U minh Forest)

For years after the Vietnam War was over, the 1973 Cease Fire was one indication of the end of the Vietnam War. However, there was evidence that U.S. POWs were still being held in the U Minh Forest Area and enslaved. During the war, the U Minh Forest, an area of thick mangrove swamps located in the northwest quadrant of the Ca Mau Peninsula, was a feared VC stronghold and the site of more than forty confirmed communist prisons and detention facilities. After the U.S. Vietnam War ended in 1974, reports reaching the Defense Intelligence Agency (DIA) in the late 1970s and early 1980s indicated the Communists were not only holding South Vietnamese prisoners in the U Minh but American POWs as well.

An example is Joint Rescue Casualty Center (JRSC) Case #3115: "The Western Caucasian Prisoners Rumored to be American POWs Seen Being Escorted into the U Minh Forest (Palm Forest of Darkness) in Southern Vietnam in 1978."

On 10 May 1978, a former South Vietnamese student who had fled to Malaysia came back to visit the Mekong Delta with a friend who worked on a collective farm near a canal on the western border of the U Minh. They noticed a group of eight men disembarking from a single sampan. The former student said that six men were "western Caucasian prisoners," and two were Vietnamese guards armed with AK-47 rifles. He said that as he and his friend watched from approximately 200 meters. The group got out of sampan and headed east on foot into the brush. They moved into the U Minh and disappeared. He said that the rumor among the laborers at the collective farm was that the prisoners he and his friend had seen were "former U.S. POWs that had been captured before 1975." Other sightings of U.S. POWs in and around the U Minh Forest are numbered 1 to 10, as described in Figure 10. JRSC dismissed Case #3115. It remains one of the extensive coverups following the Vietnam War.

1. Three American POW's were reportedly seen in a "prohibited zone" in the upper (northern) U-Minh in August 1975 as they were being escorted along a trail by two Vietnamese guards.

2. Two White Americans, both later described as very thin and with "heads shaven like monks," were seen at a dock at Xeo Ro (vic. Rach Gia), just north of the U-Minh, in October 1975.

3. In December 1975, two American prisoners were seen in prison camp in northern reaches of the U-Minh.

4. Two Caucasian prisoners were seen entering a prison just northeast of the U-Minh in Sept/Oct 1976.

5. In mid-1977, U.S. officials learned that a recent escapee from Vietnam had reportedly seen "a number of Americans, both Blacks and Whites being detained in a labor camp ... in a very remote area of the U-Minh forests.

6. In 1978, the Special Office had received another, similar report of Americans being held in a labor camp environment deep inside the lower (southern) U-Minh.

7. In 1979, U.S. officials learned of another sighting in the U-Minh in the summer of 1978, this one of two bound Caucasian prisoners and three guards reportedly seen in a motorized sampan in the central U-Minh.

8 & 9. Additional reporting from other sources told of American prisoners being held in the central U-Minh and still others at a camp near Xom Chi Chanh on the Forest's northeastern border.

10. Authors' color map entitled "Reported Sightings of U.S. Prisoners In and Around the U-Minh Forest 1975-1978.

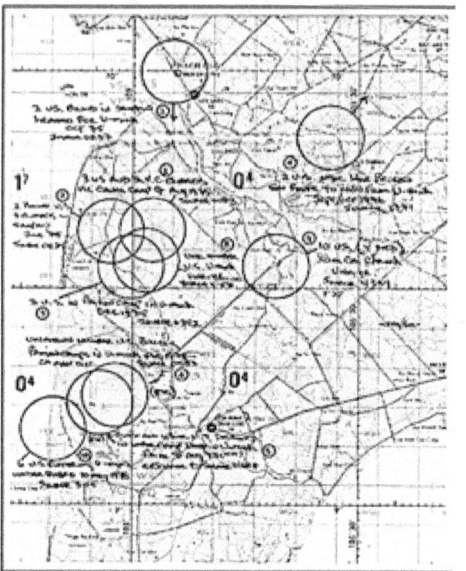

Figure 10. Authors' color map entitled "Reported Sightings of U.S. Prisoners In and Around the U-Minh Forest 1975-1979.

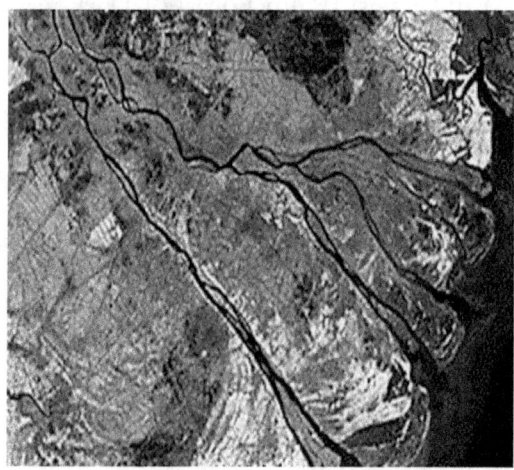

Figure 11. Aerial Map of the Mekong River Delta

Other smaller mangrove swamps were throughout the Mekong Delta, especially along the Mekong River tributaries. The Bassac, Co Chen, Ham Luong, and My Tho rivers flow into the South China Sea, as illustrated in Figure 11 aerial map. The Mekong River is the tenth-largest river in the world. The Mekong River Basin drains a total land area of 795,000 km² from the eastern watershed of the Tibetan Plateau to the Vietnamese Mekong Delta. The Mekong River flows approximately 4,909 km through three provinces of China, continuing into Myanmar, Laos, Thailand, Cambodia, and Vietnam before emptying into the South China Sea described in Figure 10. JRSC dismissed Case #3115. It remains one of the extensive coverups following the Vietnam War.

Other smaller mangrove swamps were throughout the Mekong Delta, especially along the Mekong River tributaries. The Bassac, Co Chen, Ham Luong, and My Tho rivers flow into the South China Sea, as illustrated in Figure 11 aerial map. The Mekong River is the tenth-largest river in the world. The Mekong River Basin drains a total land area of 795,000 km² from the eastern watershed of the Tibetan Plateau to the Vietnamese Mekong Delta. The Mekong River flows approximately 4,909 km through three provinces of China, continuing into Myanmar, Laos, Thailand, Cambodia, and Vietnam before emptying into the South China Sea. The Mekong River Basin includes seven broad physiographic or geomorphologic regions.

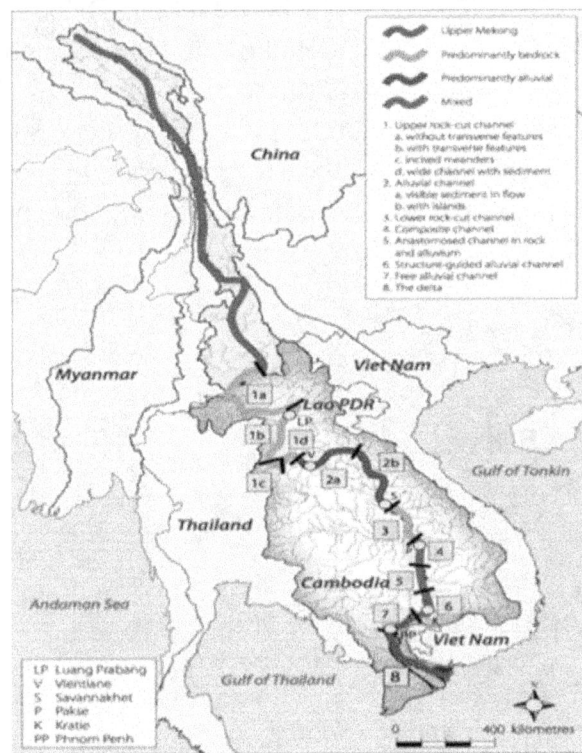

Figure 12 Mekong River Basin

They are illustrated in Figure 12, featuring diverse topography, drainage patterns, and geomorphology. The Tibetan Plateau, Three Rivers Area, and Lancang Basin form the Upper Mekong Basin (UMB). The Northern Highlands, Khorat Plateau, Tonle Sap Basin, and Mekong Delta make up the Lower Mekong Basin (LMB). (Mekong River Basin, Geography). Forty percent of the South Vietnam Republic's population of sixteen million calls this rich delta rice bowl home. Since ancient times, the Mekong Delta has been a coveted prize tempting emperors, despots, warlords, and soldiers of fortune from both Asia and Europe. So, it is not surprising that with war raging in this troubled land, nowhere was the contest more bitter or fierce than in the Mekong Delta, the breadbasket of Southeast Asia.

Figure 13. The Plain of Reeds

The Plain of Reeds in the northern Mekong Delta is close to the Cambodian border and Parrot's Beak, Figure 13, and is the shortest distance to Saigon. It was another critical axis route by VC and PAVN forces. The Plain of Reeds is an inland wetland in Vietnam's northern Mekong Delta, as identified in Figure 3 and illustrated in Figures 13.

The 162nd Assault Helicopter Company (AHC), and other 13th CAB companies, had many operations out of the Moc Hoa Airfield, and its proximity to the Parrot's Beak, Figure 14. Some of these operations are described in the stories in this chapter.

Figure 14. Moc Hoa and Parrot's Beak Location

Another key U.S. military unit successfully operating in the Mekong Delta was the Mobile Riverine Force (MRF). By 1966, the Viet Cong was prosecuting over 1,000 small-scale attacks per month on government posts and isolated villages in the Mekong Delta. The MRF grew out of General Westmoreland's desire to reduce these attacks by destroying main force units operating there. The MRF, in short, was conceived as a

Figure16 MRF Armored Transport Carrier (ATC)

means of projecting ground power into a swampy delta interlaced with waterways and rice paddies. Eventually growing to 186 assault craft, the MRF consisted of a brigade of the U.S. Army's 9th Infantry Division and a Navy component called Task Force-117 — two river assault squadrons, each with forty-five modified landing craft and other boats. The primary boat of the MRF, the Armored Transport Carrier (ATC), shown in Figure 16, was a 56-foot-long landing craft mechanized (LCM) variant. Steel and bar armor provided ballistic defense for rounds up to .50-caliber and protection against high explosive antitank rounds. Equipped with a 20mm cannon plus machine guns, the boat could travel up to six knots, fully loaded with forty soldiers. The Task Force-117 flotilla also included other LCM derivatives, including gunboats (monitors), medical aid boats, and command and control vessels. Helicopters provided an

Figure 17. Navy Sea Wolf supporting MRF

essential capability for the MRF, including the U.S. Army's 9th Infantry Division. The MRF mobility was greatly enhanced by the Navy Sea Wolves with their UH-1B helicopters, as illustrated in Figure 17, and the 164th CAG 13th CAB with its lift ships and gunships, e.g., UH-1D/H lift ships and UH-1C gunships.

19

Another critical element of the MRF and other operations in the Mekong Delta was the U.S. Navy SEAL (Sea, Air & Land) teams. They served in Vietnam between 1962 and 1972, primarily in the Mekong Delta. Operating in seven-man teams, they were typically well camouflaged and carried tremendous firepower. In addition to their offensive operations, SEALs also trained and advised their Vietnamese counterparts, the Lien Doc Nguoi Nhia (LDNN), and supported LDNN coastal missions in North Vietnam. The SEALs' fearsome appearance and extraordinary combat success prompted the Viet Cong to nickname them "Devils with Green Faces." Three Navy SEALs were awarded the Medal of Honor for their actions in Vietnam: Joseph Kerrey, Thomas Norris, and Michael Thornton.

Figure 18. Navy SEAL Teams Operating from Helicopter in the Mekong DeltaFigure

An example of typical Navy SEAL Teams operating from helicopters in the Mekong Delta is in Figure 18. Stories of 13th CAB support of Navy SEALs in search for U.S. POWs are provided Under the Phoenix Program, there were attempts by Navy SEALs working with Chieu Hoi, converted Viet Cong prisoners, to find and free the approximately one dozen U.S. POWs held in the Mekong Delta. They used Navy Sea Wolf and 13th CAB helicopters to surprise the Viet Cong POW camps to free U.S. POWs. A story of several close attempts will be described in My Stories in this chapter.

SEA FLOAT/Solid Anchor, Ca Mau Peninsula, An Xuyen Province. It was called Operation Sea Float/Solid Anchor by the U.S. Navy and Tran Hung Dao III by the South Vietnamese. Operation Sea Float/Solid Anchor was a joint U.S./Vietnamese attempt to inject an allied presence into An Xuyen Province, 175 miles southwest of Saigon.

Figure 19. SEA FLOAT/Solid Anchor on Cau Ca Mau Peninsula

Figure 20 SEA FLOAT/Solid Anchor on Cau Ca Mau Peninsula

It is illustrated in Figure 19 and Figure 20. Its purpose was to extend allied control over the strategic Nam Can region of the Ca Mau peninsula. Heavily forested, the area sprawled across miles of mangrove swamps. The site selected was on the Cau Lon River, connected to the Bo De and Be Hap rivers. These were saltwater rivers, and any fresh or drinking water used afloat or ashore had to be brought in by ship. The entire area had been solidly held by the Viet Minh against the French and by the Viet Cong against the Saigon government (and its American ally) (Vietnam Studies, Riverine Operations, 1966-1969, U.S. Army)

A Brief History of the 162nd AHC and My Initial Assignment

Upon arrival in Vietnam on February 3, 1966, the 162nd AHC was sent to Phouc Vinh, just north of Saigon, and was assigned to the 11th CAB at Phu Loi, which supported the First Infantry Division in the III Corps area. On November 1, 1968, the Company was relocated to Dong Tam in the upper Delta region and assigned to the 214th CAB, the direct support aviation battalion for the 9th Infantry Division. Upon the withdrawal of the U.S. 9th Division from Vietnam in August 1969, the 162nd and the 191st AHCs were assigned to the 13th CAB in the Mekong Delta and moved to Can Tho. The facilities at Can Tho were a few steps down from those at Dong Tam. The 191st AHC Boomerangs were stored in wooden barracks, and the 162nd AHC Vultures in tents. However, they were given the new and more powerful UH-1H helicopters, while the 191st AHC had to use the underpowered UH-1Ds.

Instead of wood, 2-story barracks, the unit lived in GP Medium tents for over a year while permanent facilities were built. Living in a tent was not ideal, especially in the Delta. Dirt, dust, and mud (in the rainy season) were everywhere, as were the mosquitoes and bugs. The mosquito nets quickly became clogged with dust, and the fine dust was constantly in and on everything. In addition, rats as large as cats and dogs would sometimes come into the "hooches" even in broad daylight. Many of the night crews slept in the daytime and kept a pistol handy to deal with any rats that came into their hooch. Once during my time with the 162nd AHC, one of our pilots was about halfway across the grassy area to the showers when he saw a large green snake slither quickly from beneath his flip flop "protected" feet. He froze in terror as he saw what clearly looked like a Green Mamba, one of the most poisonous snakes. The Green Mamba can be very aggressive, and there was no anti-venom for their bite. He said he only had two fears in Vietnam, getting bit by a poisonous snake and getting captured by the enemy. Some 162nd folks felt they lived like gypsies and wondered who the Vultures had "ticked off" to deserve such conditions. A picture of me in the 162nd AHC Tent City is in Figure 21. After the new wooden barracks were built, they were given

Figure 21. CPT Schrage in the 162nd AHS Tent

to the CH-47 Inn Keepers, and the 162nd AHC continued to live in a tent city. Figure 21. CPT Schrage in the 162ⁿᵈ AHS Tent Area

The period at Can Tho saw a shift in the type of missions flown from the previous emphasis on Combat Assaults (CAs) and related support to more ash and trash type missions with fewer CAs and a variety of sometimes unusual tasks, such as hunter-killer missions out of Moc Hoa and U.S. Embassy Missions in Phnom Penh, Cambodia in unmarked helicopters. After the pullout of the 9th Division (parts of the Division did not leave until early 1970), there were no U.S. infantry units left in the Delta. The Vultures and other AHCs then primarily supported ARVN units. The ARVN units included the 7th, 9th, and 21st ARVN Infantry Divisions and other Vietnamese and Navy SEAL units throughout the IV Corps area until the Company stood down in April 1972. The 162nd had the distinction of being the last assault helicopter unit in the Delta. The primary company patch for the 162nd AHC was the Vulture for the Company, and the two lift platoons, and a Copperhead for the gunship platoon, as illustrated in Figure 22.

Figure 22. The Primary Patches for the 162nd AHC

Two of us, an Infantry Officer Captain Brian Bush, and myself, a Field Artillery Officer Captain, both directly arriving from Flight School, were assigned to the 162nd AHC. After meeting with the 162nd AHS

Figure 23A. CPT Schrage leaving his tent area

Figure 23B. Vulture 16 AC

Figure 23C. Combat Assaults

Commander, MAJ Ken Loveless, I was appointed as the First Lift Platoon Leader, Vulture 16. CPT Bush was appointed the Second Lift Platoon Leader, Vulture 26. There were only four captains in the 162nd AHC, of which we were two. CPT George Hawkins, the Executive Officer, and CPT Walt Stewart, the Gunship Platform Leader, were highly respected for their experience in successful air mobility combat operations. The rest of the pilots were seasoned warrant officers, sprinkled with a few lieutenants. Even though we were "newbies," most of the warrant officers were experienced, had served several tours in Vietnam, and some even married South Vietnam women while doing continuous tours. They were also the Senior Aircraft Commanders (ACs), while we still had to earn and obtain this rank through check rides and demonstrate our flying skills in combat.

Illustrated in Figure 23 A&B is CPT Schrage, an AC, leaving his tent area for his Vulture 16 aircraft for a combat assault mission, as shown in Figure 23C.

Experiences and Stories with the 162nd AHC from February- July 1970

My Promotion to Aircraft Commander and Company Air Mission Commander

I was successfully promoted to AC in March 1970 and spent most of my time flying my UH-1H and leading combat assaults. Each of the AHCs had to field "a package" daily, which consisted of five lift ships, two gunships, and a command and control (C&C) helicopter. A typical day began with a wakeup at 4:30 a.m. and a quick breakfast. This was followed by an aircraft pre-check, then departing with or leading "the package" by 6 a.m. Often this was heading for the location that needed combat support. The Air Mission Commander (AMC) was usually the 162nd AHC Commander, MAJ Ken Loveless, his Executive Officer, CPT George Hawkins, or possibly one of the Lift Platoon Leaders, also ACs. With CPT Hawkins rotating back to the U.S. and MAJ Loveless becoming a "short-timer," I soon became an AMC from April-July 1970 before being assigned as the Assistant S-3 for the 13th CAB in Soc Trang. I served as the AMC on a several-day encounter with a PAVN Division moving out of the Seven Sister Mountains Region, crossing over to the U Minh Forest. This crossing over-involved providing air assets from the 13th CAB, the 307th CAB, and a VNAF Company to support the 21st ARVN Division. I served as the overall AMC and logged 12 hours of flight time the first day and 13 hours the second day. Over my three months with the 162nd AHC, I logged over 700 flight hours and received thirty-five Air Medals during 1970. When it got dark the first day, the VNAF package left the battle, stating, "VNAF go home now," as they didn't fly in combat at night. The respect between the ARVN and VNAF was not good and indicated that Helicopter Vietnamization would be difficult to achieve. Later, when I became the S-3, 13th CAB, the 21st ARVN Operations Center requested combat air support. They stated, "No VNAF," and they would prefer getting "Dark Horse," C Troop, 16th Air Cav, the 121st AHC Soc Trang Tigers or the 336th AHC Warriors.

PARTICIPATION IN THE CAMBODIA INCURSION AND THE LARGEST AIR MOBILE ASSAULT

A change in the Cambodian government allowed an opportunity to destroy the PAVN bases in Cambodia in 1970 when Prince **Norodom Sihanouk** was deposed and replaced by pro–U.S. General **Lon Nol**. The Cambodian Campaign (also known as the Cambodian Incursion and the Cambodian Invasion) was a brief series of military operations conducted in eastern Cambodia in Spring 1970 by South Vietnam and the United States forces as an extension of the Vietnamese and Cambodian Civil Wars.

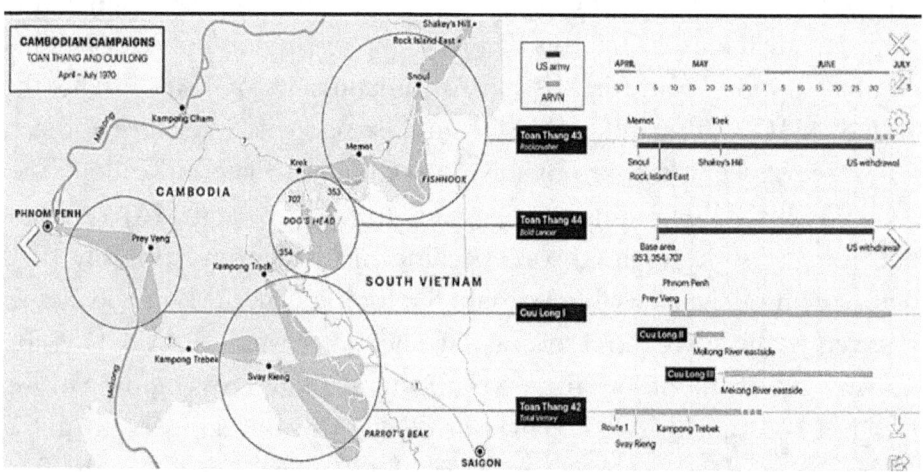

Figure 24 Thirteen Major Operations in Cambodia Incursion with Arrows Identification of Objectives in Spring and Early Summer 1970

As illustrated in Figure 24, the Army of the Republic of Vietnam (ARVN), between 29 April and 22 July, and U.S. forces between 1 May and 30 June, exercised thirteen significant operations. The Campaign's objective was the defeat of the approximately 40,000 troops of the People's Army of Vietnam (PAVN), The Khmer Rouge, and the VC in the eastern border regions of Cambodia. Under Prince Sihanouk, Cambodian neutrality and military weakness made its territory a safe zone where PAVN/VC forces could establish bases for operations over the border. With the U.S. shifting toward a policy of **Vietnamization** and withdrawal, it sought to shore up the South Vietnamese government by eliminating the cross-border threat.

However, the incursion into Cambodia was too short, especially considering General Lon Nol's intent to convert Cambodia into a democracy. He urgently requested the U.S. for assistance, including helicopters, in developing a democracy and ridding Cambodia of PAVN/NVA forces while defeating the Khmer Rouge Communists in the northern part of Cambodia. A coalition of Cambodia, South Vietnam, and U.S. Forces could have created a democracy and halted the spread of Communism-the original goal of the Vietnam War. Yes, this Coalition could offer the opportunity to "Win the Vietnam War".

Unfortunately, the U.S. didn't respond to General Lon Nol's urgent request for helicopters. A viable coalition of Cambodia and South Vietnam with U.S. support could probably have provided a combined military defense with new Borders boundaries 1 or 2, as illustrated in Figure 25A. Proposed Border 1 Boundary would be preferred. It could include the Southern portion of Cambodia including Phnom Penh and in South Vietnam the Mekong Delta plus Saigon and Bien Hoa, as shown in Figure 3; in essence the IV and III CTZs. Outcomes of a joint Border 1 Boundary Coalition could potentially have produced the following results for both Cambodia, South Vietnam and the U.S.

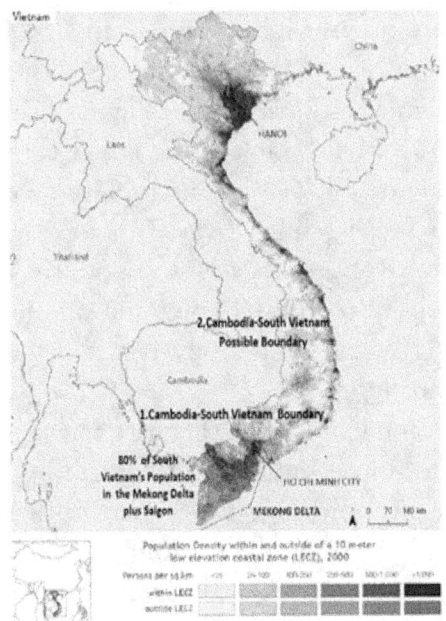

Figure 25A. Vietnam Population Distribution. *Figure 25B. Infinite # of Infiltration Routes*

A Border 1 Coalition could possibly have provided many of the following results:

1. The inclusion of the cultural and breadbasket areas of each country.

2. The inclusion of approximately 60-80% of the populations in both countries in Figure 25A

3. The inclusion of the CTZ IV and CTZ III military bases and operations areas in South Vietnam

4. The elimination of many of the infinite infiltration routes into South Vietnam in Figure 25B. This was never possible with the length of the country and Ho Chi Minh and Sihanouk Trails across the border along with easy access from the South China Sea.

5. In hindsight, it could have resulted in "Winning the Vietnam War" and satisfy the U.S. Goal of halting the spread of Communism.

6. In hindsight, it could also have eliminated much of the genocide in Cambodia, as described in The Killing Fields, and freed the remaining U.S. POWs in South Vietnam.

The primary objective of the Cambodia Incursion was disruption of the supply chains and destruction of PAVN and Viet Cong forces along with the Ho Chi Minh and Sihanouk Trails, as illustrated in Figure 24B. In the Mekong Delta, a helicopter Air Mobile Assault of the 9th ARVN Division, shown by the large arrow in Figure 26, was sent up the Mekong River to free up the Highway 1 River Crossing. The River Crossing was taken over by the PAVN and the Khmer Rouge forces to stop both river and highway traffic. A total of 2425 troops were inserted into nine landing zones in the vicinity of the Ferry Site crossing of Highway 1. It was the largest helicopter airmobile assault in history.

The Assault was followed by Operation Cou Long I, Figure 24, which commenced with a four-pronged assault into Cambodia. Two armored columns and a naval column crossed the border and hastened up the east side of the Mekong River to link up with the Air Mobile Assault Force. Each AHC was tasked with providing a package of five lift ships, two or three gunships, and a command and control ship. The total Air Mobile Force of sixty lift ships could airlift almost the entire 9th ARVN

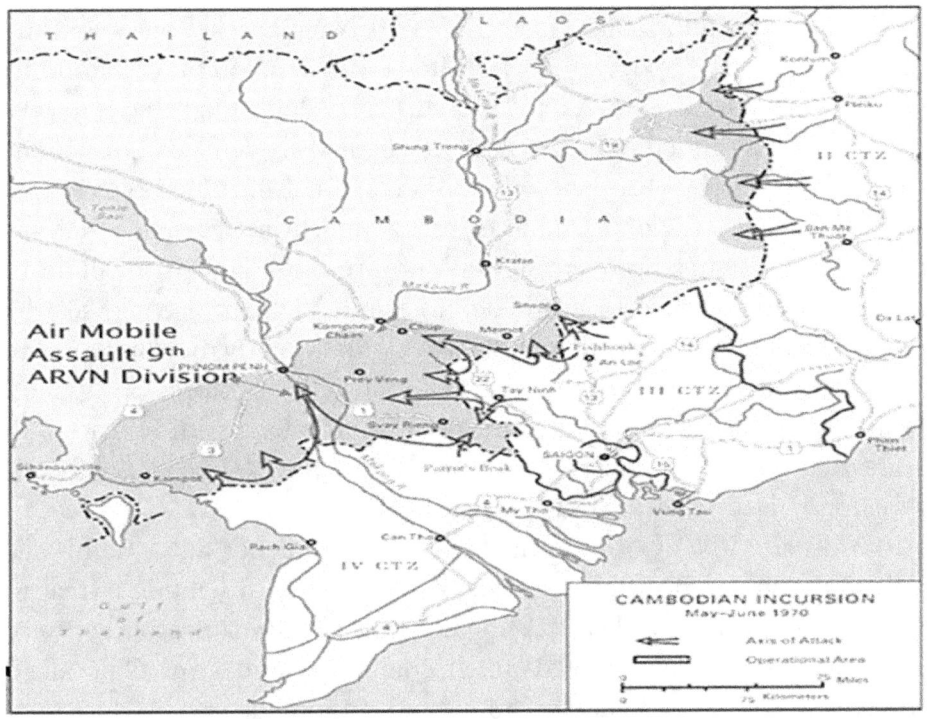

Figure 26. Helicopter Air Mobile Assault

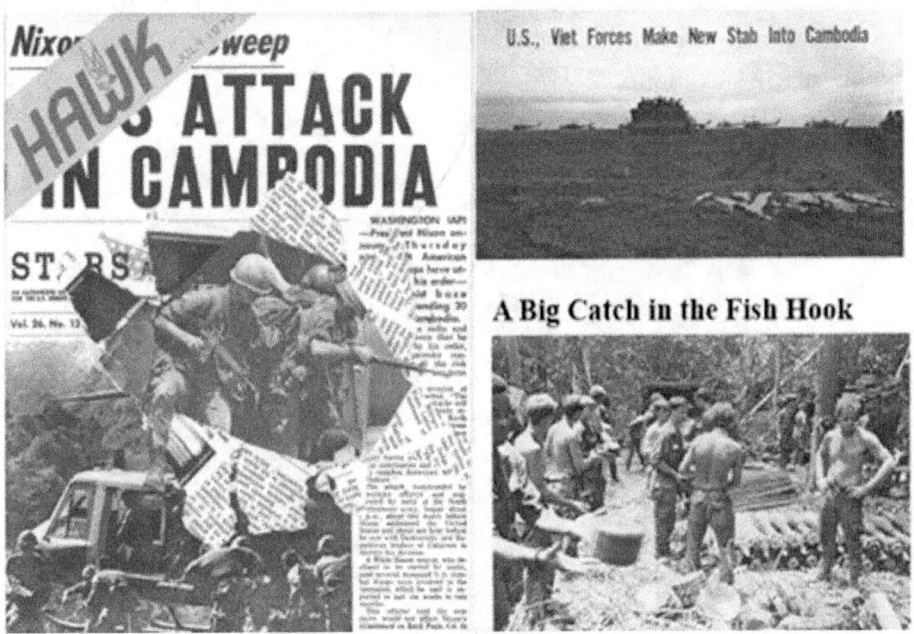

Figure 27. Combat Operations in Cambodia and Ammunitions & Supplies Found

Division. Air support came from the A-10 Navy Black Ponies and the Air Force C-130 "Spooky" assets. After refueling along the Cambodia Border, the Air Mobile Assault Force proceeded along the extensive arrow route illustrated in Figure 26 at first light in sixteen Vee Formations of five UH-1D/H helicopters one minute apart. I flew the point in one of the first Vees. We expected to be met at the River Crossing by PAVN air defense systems; however, we took no fire as we landed. I will never forget when we turned to head back to South Vietnam. I could see the perfect flight formations of the other Vees with the early morning sunlight rising over the horizon. In the days that followed, vast stores of supplies, ammunition, and equipment were discovered, along with contact with some of the PAVN troops, as illustrated on the July 1970 cover of the 1st Aviation Brigade Hawk Magazine, Figure 27.

Most PAVN/VC forces had withdrawn deeper into Cambodia before the invasion, with a rear guard left to stage a fighting retreat to avoid charges of cowardice. PAVN/VC losses in human resources were minimal, but they abandoned much equipment and arms. The allied forces captured a vast haul of weapons and equipment, and for the rest of 1970, notably, reduced PAVN/VC activities in the Saigon area.

Most of PAVN/VC forces were instructed by their leaders to withdraw to north Cambodia and not engage the ARVN-US Forces, as illustrated by the arrows in Figure 27B.

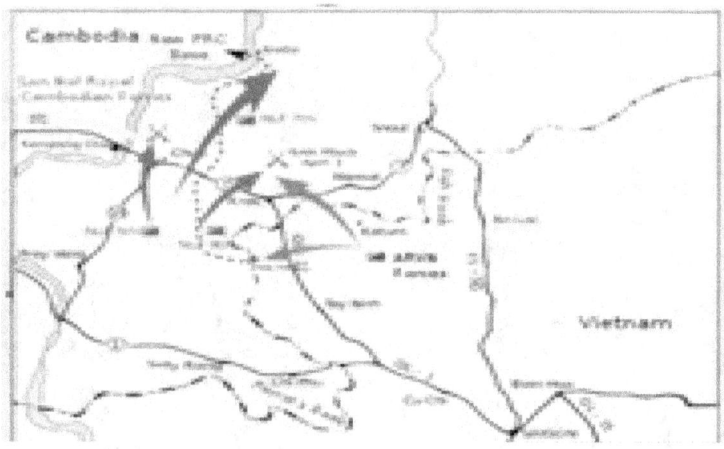

Figure 27B. PAVN/VC Forces Movement

When General Abrams was informed of this PAVN/VC retreat, he saw that an incomparable opportunity was at hand. The constraints on the operation in terms of time and depth were agonizing to Abrams as a soldier. However, he concluded the following actions were required but took no action. (Sorley, A Better War, Cambodia Chapter)

- "What we need right now is another division –go in deep," he said in the wake of the initial penetrations.

- "We need to go west from where we are, we need to go north and east from where we are. And we need to do it now. It's moving and –goddamn, goddamn"

- This last was said wistfully, with great sorrow and regret. Abrams said "Time to exploit."
"Christ – it's so clear. Don't let them pick up the pieces. Don't let them," he said very softly, "pick up the pieces. Just like the Germans –you give them thirty-six hours and, goddamn it, you've got to start the war all over again."

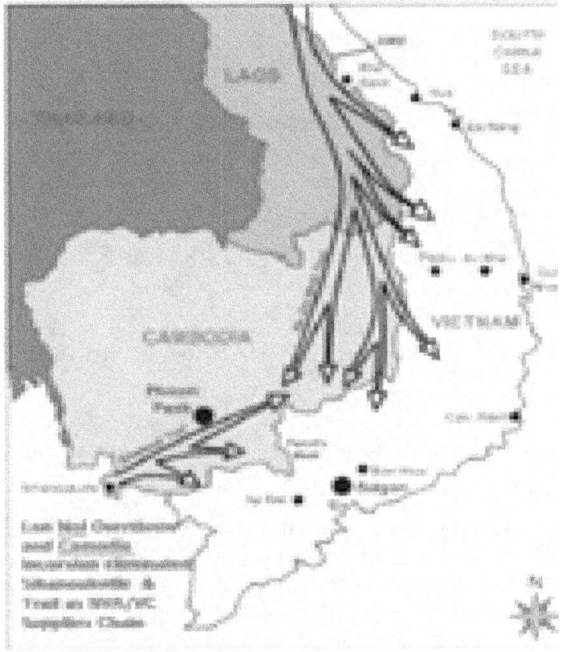

Figure 27C. Shutdown of Sihanoukville & Trail

31

However, no action was taken by General Abrams, even though he was the MACV military commander of the Vietnam War. There was other divisions available in the Mekong Delta, The 21st ARVN Division, probably the best ARVN Division in Vietnam supported by the 13th CAB. It had participated along with the 9th ARVN Division in Cuu Long I airmobile assault and returned to the Mekong Delta. With the 164th CAG available airmobile support the ARVN could have pursued the PAVN/VC in Cambodia in June-July 1970. This could have leveraged the Lon Nol government actions which shut down Sihanoukville and most of the Sihanouk Trail, Figure 27C, in Cambodia. For the potential Cambodia, South Vietnam and US coalition this was probably the best and last opportunity for combined ARVN, FANK (Khmer Republic Armed Forces [Cambodia]) and US forces to sustain Government of Cambodia (GOC) and "Win the Vietnam War". Wars aren't won without boots on the ground and airmobile and gunship support.

- Airmobile Operations will be essential for future warfare and must be included as Direct Combat and Integral Combat Support vice General Support

- Boots on the Ground do not have to be US troops if the ground force is properly trained and disciplined, such as the 21st ARVN Division and 9th ARVN Division in the Mekong Delta

- While airmobile operations in Vietnam were at the tactical level and provided unit support at the battalion and company ley vel, can this be sustained in the future?

- LTG John J. Tolson's 1961-1971 Airmobility Study, 1999 should be reviewed.

- GEN Howze concluded from a 1966 visit to Vietnam that the US would have been subject to defeat just as the French – without the Helicopter. This is why General Lon Nol wanted helicopters for Cambodia.

MY COMBAT MISSIONS STORIES IN CAMBODIA
AND IMPACT ON VIETNAMIZATION

The Combat Operations into Cambodia in April-May-June 1970 were intended to clear out supplies, ammunition, and forces along the Ho Chi Minh and Sihanouk Trails to give the U.S. Vietnamization effort a chance for success. However, by 1971 the PAVN/VC had replaced all weapons and equipment. The PAVN/VC also returned to their frontier bases in the summer of 1970, after the Americans' withdrawal in June 1970. In the four months after Sihanouk's ouster, the Communists had overrun half of Cambodia, taken or threatened sixteen of its nineteen provincial capitals, and interdicted—for varying periods—all road and rail links to the capital, Phnom Penh. In the countryside, VC/NVA forces generally continued to move at will, attacking towns and villages in the south and converting the north into an extension of the Laos corridor and a base for "people's war" throughout the Country and South Vietnam. This being the situation, the survival of the Lon Nol government would depend heavily on the extent of foreign assistance, and the will and ability of the people. Their leaders needed to organize themselves for effective military resistance to the Communists on both the unity and morale of the country in the face of hardship, destruction, and death; and on the reaction to the divisive political appeals issued in Sihanouk's name.

But of equal or greater importance was the capabilities and intentions of the Vietnamese Communists in Cambodia, the extent to which they could bring pressures to bear on the Lon Nol government, and the degree to which they were willing to allocate available resources to such an effort. (*Foreign Relations, 1969–1976, Volume VII. Vietnam July 1970-January 1972.* Special National Intelligence Estimate 1 SNIE 57–70, was released in SEPTEMBER 2010)

One mission that the 162nd AHC had in Cambodia, even before the Incursion, was supporting the U.S. Embassy in Phnom Penh. It consisted of flying an unmarked UH-1H/D lift ship with the crew in civilian clothes. I flew two of these missions. One before the Incursion and one after the Incursion. The first mission was enjoyable as we transported representatives of the U.S. Embassy and spent the night in an excellent French Hotel, which reflected why Phnom Penh was called "the Paris of the East."

After the Incursion, the second mission was when the PAVN and Khmer Rouge had begun encroaching on Phnom Penh. As we approached Phnom Penh Airport, we could not establish radio contact with the control tower. There was a DC-3 Aircraft blown up on the end of the runway, and hangars, which housed the few Cambodian helicopters, had 122 mm Mortar holes in them, destroying most of the Cambodian Air Force. The next day we flew missions with U.S. Embassy personnel to determine which outlying villages around Phnom Penh were under whose control, i.e., the Cambodian Khmer Army or the PAVN or Khmer Rouge. I was glad to fly back to Vietnam as soon as possible and realized that a push by the Khmer Rouge and PAVN to take Phnom Penh was beginning.

This reality was further reinforced when I flew an Air Combat Mission into Cambodia. This mission provided command and control for helicopter gunships (UH-1Cs) and attack helicopters (AH-1Gs). They were trying to support the Cambodia Khmer Army in keeping Highway 4 open through the Elephant Mountains, between Phnom Penh and Sihanoukville on the coast, as illustrated in Figure 28.

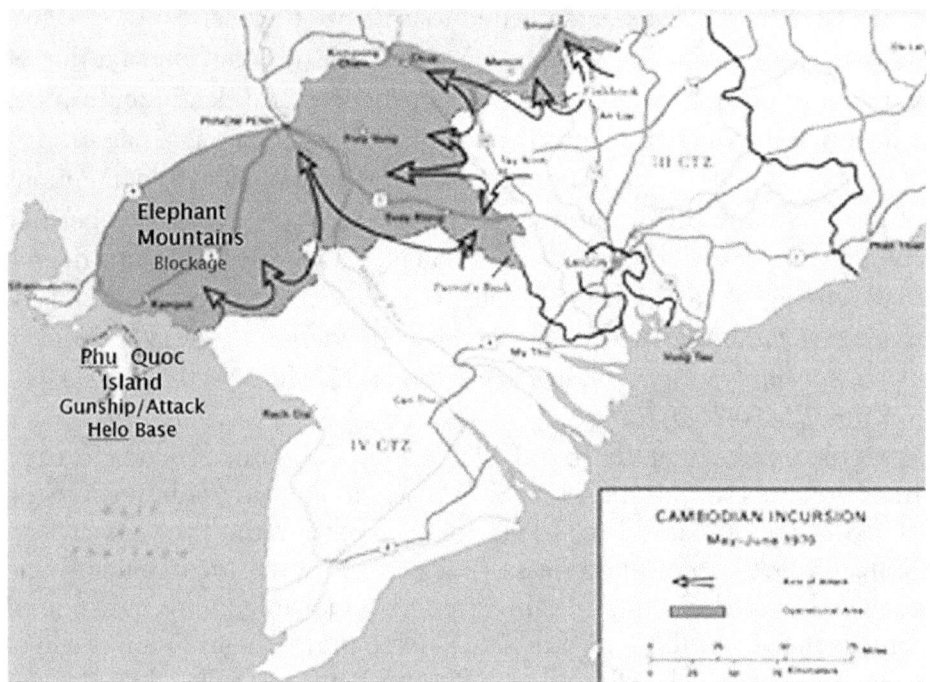

Figure 28. Cambodia Combat Operations off Phu Quoc Island

The gunships and attack helicopters from the 13th CAB and the 307th CAB had to be stationed on Phu Quoc Island to have the necessary range to support the Cambodia Khmer Republic Army in trying to disrupt the Khmer Rouge/PAVN blockage in the Elephant Mountains. Radio contact with the Cambodia Khmer Republic Army ground troops was problematic, often having the gunships and attack helicopters returning to Phu Quoc Island without firing. I landed my command and control aircraft near a shrine in Cambodia. I met with the Cambodia Khmer Republic Army leaders to discuss better communications with the gunships and attack helicopters. I was shocked by the age of some of the Khmer troops, who were 14-15-year-old kids. I got the feeling they felt the U.S. was going to come in and save them, similar to Vietnam, although I knew that our Vietnamization efforts did not include Cambodia.

The Missed Opportunity was for a Better War in a Defendable Region through a Coalition of Southern Cambodia-South Vietnam Delta, CTZ IV, and Saigon-Binh Hoa CTZ III

This missed opportunity for a Cambodia-South Vietnam Coalition was introduced earlier in Chapter 2 when describing Participation in the Cambodia Incursion and the Largest Air Mobile Assault. It will now be discussed to set the record straight. This discussion regards the primary Book 1 question: Why was winning the Vietnam War unsuccessful? The book, *A Better War-The Unexamined Victories and Final Tragedy of America's Last Years in Vietnam*, by Lewis Sarley, published in 1999, sheds some light on answering this question. It is based on the Abrams Papers or Abrams Special Collection of General Creighton W. Abrams, commander of U.S. forces in Vietnam from 1968-72. It explains why" A Better War" could have been successful and describes the change in approach when General Abrams took over as commander of U.S. forces in Vietnam from General William C. Westmoreland following the Tet '68 Offensive.

A review of the following on failed U.S. nation-building efforts in Iraq and Afghanistan shows that the U.S. still has not learned how to apply the *clear, hold* and *build* functions named in the Sorley book.

Figure 29 Three Competent ARVN Divisions

In 1970, the 7th, 9th, and 21st ARVN Divisions identified in Figure 29 had proved themselves in battle and progressed well under the Vietnamization Program.

Figure 30. 164th Combat Aviation Group

The 164th CAG, shown in Figure 30, identified and applied the necessary airmobile support and firepower following Tet '68 for expanding complete control of the Mekong Delta and beyond. With the GEN Lon Nol Overthrow of the Prince Sihanuak Regime in Cambodia in early 1970, it is believed that a successful coalition of two united countries formed from southern Cambodia, including Phnom Penh, and from the Mekong Delta, including Saigon and Bieh Hoa, along with most of CTZ III, could probably have provided *A Better War* solution and possibly won the Vietnam War.

It can be seen in Figure 31A, and it is known that the Tet '68 NVA and Viet Cong attacks were repelled, and the U.S. and South Vietnam won in the Mekong Delta, IV CTZ, and III CTZ. They also later won in III CTZ with ARVN troops and 164th CAG air support during the 1972 Easter Offensive, Figure 31B.

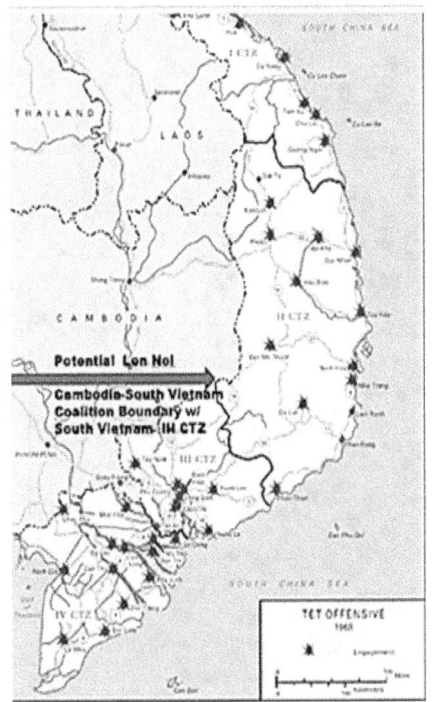

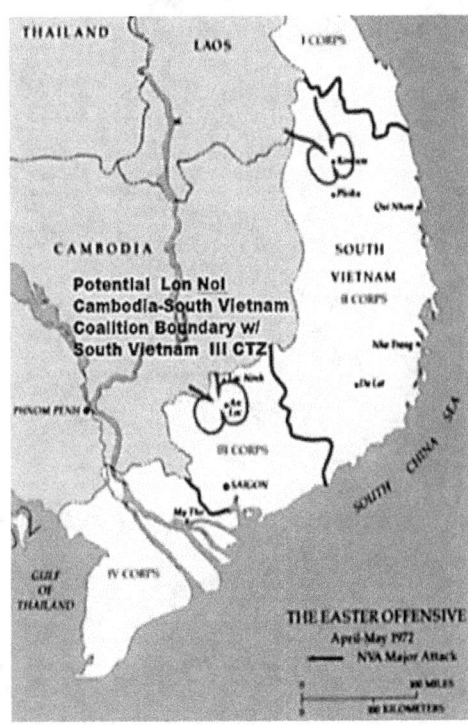

Figure 31A. Tes'68 Attacks on South Vietnam   *Figure 31B. A Map of the 1972 Easter Offensive*

Following the Vietnam War, information on the importance of the Mekong Delta to the enemy was in a quote obtained from the NVA/ PAVN Central Office for South Vietnam's Resolution No. 9 disseminated in 1969. It *emphasized the strategic importance of the Mekong Delta*, summarized in Figure 32. The NVA/PAVN leadership conceived it as *the principal battlefield where the outcome of the war in South Vietnam would be decided.*

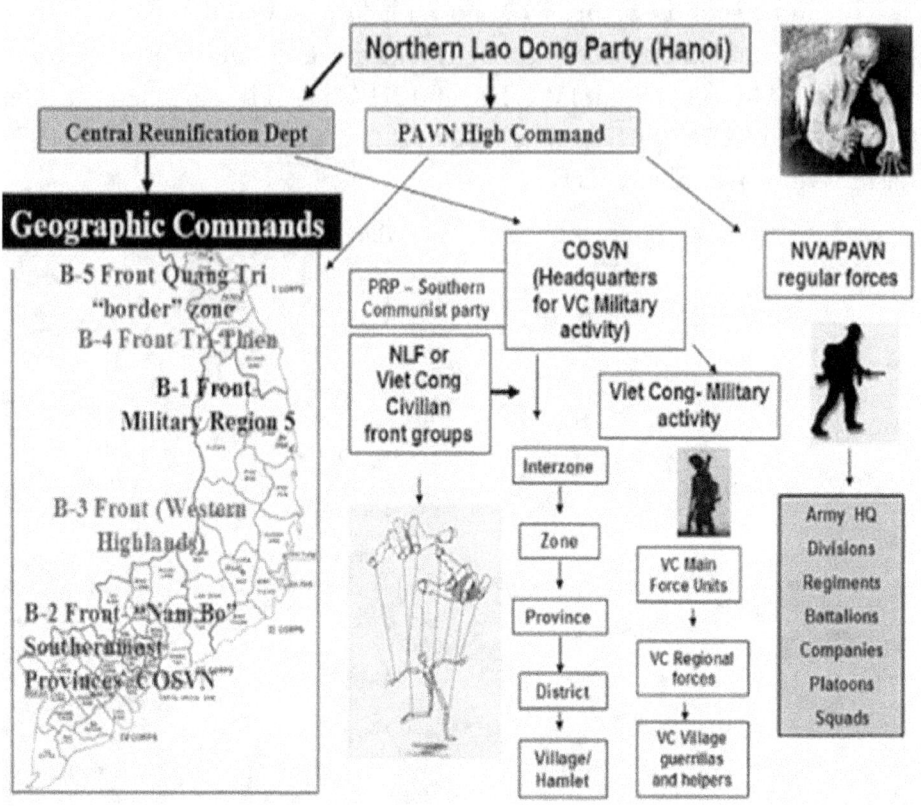

Figure 32. NVA/PAVN Central Office for South Vietnam's Resolution No. 9 disseminated in 1969

Therefore, the PAVN/VC infiltrated the 1st Division Headquarters and its three regiments, the 88th, 101D, and 95A, into IV Corps, where they were defeated or at least contained in the Seven Mountains and U Minh Forest. In hindsight, it is believed that a potential Lon Nol Cambodia, South Vietnam and USA Coalition with 164th CAG air support for the ARVN and Cambodia Forces could hold a boundary with South Vietnam, including III CTZ. This boundary, illustrated in Figure 31A, could have kept the NVA/PAVN from invading and taking over III and IV CTZ in South Vietnam. It could have kept the Khmer Rouge/PAVN from taking over Phnom Penh, possibly preventing the subsequent genocide led by Pol Pot.

HELICOPTER VIETNAMIZAITON AND TROOP REDUCTIONS BY SECRETARY MELVIN LAIRD IN FALL 1970 AND SPRING 1971.

In August 1970 President Nixon directed the establishment of a Special Review Group (SRG) for Southeast Asia comprising the under Secretary of State, the Deputy Secretary of Defense, the Director of Central Intelligence, the Chairman of the Joint Chiefs of Staff, and chaired by the Assistant to the President for National Security Affairs. The SRG prepared National Security Study Memorandum (NSSM)99. There were a total of four strategies identified in regards to supporting the Government of Cambodia (GOC) and its Relationship to Vietnamization.

—**Strategy 1:** *A Minimum Resources Strategy* (OSD's option) that deems the GOC non-essential to Vietnamization and precludes RVNAF defense of the GOC (or with Laird's caveat does not preclude RVNAF involvement but prohibits U.S. support for it.) No additional U.S. military assistance would be provided to the GOC although $45 million in available economic assistance would be provided.

—**Strategy 2:** *A Limited Resource/Involvement Strategy* (State's preference) which deems the preservation of the GOC non-essential to Vietnamization, precludes (unequivocally) RVNAF involvement to defend the GOC (as opposed to cross-border operations), but calls for an additional $100 million in U.S. economic and military assistance to give the GOC the chance to go it alone.

—**Strategy 3:** *A Defense of a Viable GOC* Strategy which uses RVNAF as necessary to defend GOC on territory ranging from one-fifth to one half of Cambodia depending on the variant chosen. This option can be defended either from a judgment that the preservation of the GOC is essential to Vietnamization success (the JCS view) or that it is beneficial to Vietnamization even with some RVNAF involvement in Cambodia (JCS Preference).

—**Strategy 4:** *Offensive Operations in South Laos and Northern Cambodia* could be conducted as part of a strategy to defend Cambodia. Logically, this option is not an alternative to the three preceding options

Strategy 3 was overall selected for implementation. However, Secretary of Defense Melvin Laird strongly supported Strategy 1. He also supported the expedition of Troop withdrawal and cut the Army funding in 1970-71

In addition to the reduction in US troop levels of 50,000 to be accomplished by October 15, 1970 Secretary Laird reduced the authorized ceiling by another 40,000 between October 15 and December 31, 1970. This expedited Drawdown by Sec Def Laird left the 13th CAB and 164th CAG unable to support ARVN Divisions in the Mekong Delta and the GOC FANK after December 1970. This directly influenced the transfer of Soc Trang AAF to VNAF and their USAF advisers which will be addressed later in this Chapter. While Vietnamization was the accepted approach and the ARVN were successful, Helicopter Vietnamization was not successful and lost the capability of the GOC, South Vietnam and US to win the war. The Secretary of Defense solution was to have VNAF helicopters to Phnom Penh to train GOC helicopter pilots which was never successful.

* * *

Operations out of Moc Hoa, and the Plain of Reeds, and my Stories

Hunter-Killer Teams, and Search and Destroy

Moc Hoa was located close to the Cambodian border, south, and west of the Parrot's Beak, which had the closest access to Saigon, as illustrated in Figures 15A and B. In 1970, the 162nd AHC had a nightly mission to fly a "search and destroy" operation out of Moc Hoa, a village and airfield along the Cambodian border in the Plain of Reeds.

Arguably, by mid to late 1969, the Mekong Delta had become a Viet Cong sideshow. Improvements in ARVN capabilities, the effectiveness of SOF programs, and the enemy miscalculation/disaster called the "68 Tet" had severely damaged enemy units and cadres. Viet Cong attacks in the Delta were down to squad and platoon size efforts and would stay that way unless the NVA intervened. The NVA chose to intervene. Their infiltration corridor was the border area between the Seven Sisters Mountains and the Parrot's Beak intersection with the III/IV Corps boundary. Their support bases were located in the Parrot's Beak itself and the area around Svey Reng in Cambodia. Thus, operations out of Moc Hoa, especially at night, were to detect PAVN infiltration across the Plain of Reeds. The 162nd AHC had this as a primary mission.

A typical mission was for one or two gunships and a C&C/flare ship to stage out of Moc Hoa. At various times during the night, they would link up with an OV-1 Mohawk aircraft and Delta Control to recon the northern portion of the Plain of Reeds. The 162nd AHC Copperheads also provided on-call support to the Special Forces outposts along the border. The Special Forces B Team was based in Moc Hoa. An OV-1 Mohawk would often pick up something on its radar and relay the coordinates to Delta Control. Delta Control would vector the Copperheads over the site. (It was here that the Copperheads learned an OV-1 was great at finding herds of water buffalo.) In any case, it was a dangerous area and a dangerous mission.

The American and Vietnamese troops who manned the string of outposts along the border had a lonely and dangerous job. As the time approached to go over the border, the Copperheads began to work around the clock to patrol from a line north of Moc Hoa west to Chau Doc and then back again. Occasionally, as far south as the Seven Mountains area, I served as the Air Mission Commander for several of the search and destroy out of Moc Hoa. For the missions that I flew, I was glad we had the UH-1C gunships rather than the AH-1G tandem seated attack helicopters. The reason was that if the command and control helicopter or one of the UH-1C gunships got shot down, they could be picked up by the other surviving cabin-based helicopters. This was something the AH-1G attack helicopters could not provide. The AH-1G attack helicopter had more speed and endurance with a tandem cockpit and could carry a heavier payload. However, it worked best in Air Cav Troops with an OH-6A Scout helicopter, i.e., Loach, in a hunter-killer scenario, operating on the deck finding the enemy, e.g., hunter, while the AH-1G attack helicopter operated from altitude, e.g. 1500 – 2000 above ground level as the killer. In the Mekong Delta, the UH-1B/C gunships with door gunner and crew ship had 360 deg protection which was more efficient for supplying combined hunter and killer capabilities.

DELIVERY OF THE CRITICAL RADAR COMPONENT TO SPECIAL FORCES

Another experience I had flying a UH-1H helicopter in the Seven Sisters Mountains Region was during the delivery of a critical radar component for one of the Special Forces Camps. I had flown up to Long Binh to pick up the component. When I landed at the Special Forces site, I had to make a vertical descent to navigate through a series of antennas, which is not a good approach as you must fly through the "dead man's zone." However, the worst case was when I left and tried to climb through the antenna field. As I climbed through the antennas, the engine RPM dropped from 6600 to 6400, indicating a weak engine. I faced a decision. I could try to nurse the aircraft up through the antenna field with increased collective pitch, which didn't seem to be working, as the engine RPM was approaching 6200, at which the aircraft wasn't supposed to fly. Or I could lower the collective and pitch the aircraft down through the antenna field. I quickly decided to lower the collective and pitch the aircraft down through the antenna field. Luckily, I made it through the antennas. This is one of two times during my tour that I felt I might not make it. I will discuss the other occasion later. When I got back to Can Tho, it was determined that the engine was underpowered.

162ND AHC COPPERHEAD GUNSHIP OPERATIONS OUT OF MOC HOA AND THE PLAIN OF REEDS

CPT Walt Stewart was one of the most respected senior aviators in the 162nd AHS during 12/69-7/70. He was serving as the Copperhead Gunship Platoon Leader when I arrived in January 1970. On many occasions, his gunship team operated out of Moc Hoa in the Plain of Reeds and close to the Parrot's Peak Cambodian Border, Figure 14. On one occasion, the Copperheads were sent out from Moc Hoa in the early morning to look for a battalion of NVA that had come off the Seven Sister Mountains. They found them, or at least they found each other! The NVA had at least one heavy machine gun and made the mistake of opening fire too soon. As CPT Stewart recalls, they expended their rockets (WO McGlamery was wing) and gave instructions to an AH-1G Cobra Team that followed them in. Unfortunately, the NVA gunners got the range on the Cobras and killed the front seat pilot in one of the aircraft.

CPT Stewart has written an excellent summary of his experiences as the Copperhead Gunship Platoon Leader and of operations out of Moc Hoa and the Cambodian Invasion. The 162nd AHC Commander MAJ Tom Beauchamp and the Copperhead Gunship Platoon selected me to be CPT Stewart's replacement when he rotated back to the States in July 1970.

During the last few weeks of June 1970, I flew with him in his UH-1C Gunship on several occasions. One particular flight was an operation out of Moc Hoa Army Airfield. CPT Stewart and I were re-arming and refueling the UH-1C Copperhead in the Moc Hoa Re-arming and Refueling point. Fully loaded UH-1C models with arms and full fuel load could not hover and often had to bouncedown the runway until achieving translational lift. Engineers had been replacing Perforated Steel Planking (PSP) matting with the new square mats on the runway. We bounced the helicopter down the runway for take-off. The rotor wash picked up a large, unsecured matting section as we tried to get off the ground. The bottom of the skids slid under the mat section just as they reached the translational lift, the cross tubes gave way, and the entire skid assembly pushed up against the bottom of the aircraft – the first and only retractable gear UH-1C. We flew the aircraft back to Can Tho, reduced fuel load, and dropped two fully loaded XM200 rocket pods in the Mekong River. We then landed on some mattresses that maintenance, crew chiefs, and pilots had set up in a revetment for a relevant soft adjusted landing. Somewhere in the Mekong River, off Can Tho, were two surprise packages, compliments of the Copperheads and the engineers at Moc Hoa. We wrote up a pilot report (PIREP) on the runway conditions at Moc Hoa. This story rapidly spread among helicopter pilots in Vietnam. Even after I returned to the States years later, David Groen, President and CEO of Groen Bros Aviation and a former Vietnam helicopter pilot, told me, "So you were one of the pilots that flew the skid-less helicopter out of Moc Hoa."

MY SELECTION AS THE ASST S-3 13TH CAB

Shortly after becoming the Copperheads Platoon Leader, I was informed by the 13th CAB Executive Officer, MAJ Louis Sokowoski, that I had been selected from a hand-picked group of officers for the position of Assistant S-3. He stated this was because of my experience as an Air Mission Commander and that my reaction to a combat environment had been outstanding. He also stated that I had obvious potential for positions of greater responsibility. I readily accepted this new assignment, based on both the reputation of the 13th CAB and the extensive knowledge I had gained with the 162nd AHC in flying missions and as an Air Mission Commander for Combat Operations in the Mekong Delta and Cambodia.

HISTORY OF THE 13TH CAB SOC TRANG ARMY AIRFIELD (AAF) AND SIGNIFICANT STORIES

The 13th Aviation Battalion was formed initially as the Delta Aviation Battalion on 4 July 1963 at Can Tho, South Viet Nam. On 5 August 1964, the 13th Aviation Battalion was activated at Fort Bragg, NC, and deployed to South Viet Nam to replace the Delta Aviation Battalion in September 1964.

The 13th Aviation Battalion made its base of operation at Can Tho and underwent a name change. It became the 13th CAB. The 13th CAB provided air mobility for IV Corp Delta Tactical Zone. In December 1967, the 13th CAB was reassigned to the 164th CAG and relocated its base of operations to Soc Trang Army Airfield in October 1968.

The 13th CAB was awarded the Presidential Unit Citation on 27 and 28 Aug 1965 for action in Chuong Thien Province. Then they received the Valorous Unit Award 4-6 Apr 65 for participation in Operation Dan Chi in support of ARVN General Dong Van Quang's 21st ARVN Division. In addition, the 13th CAB received the Outstanding Army Aviation Unit of the Year in 1964. The 13th CAB stood down in March 1972.

The Soc Trang Airfield had a rich history. Over the years, it was a French colonial, Imperial Japanese Army, United States Marine Corps (USMC), United States Army (US ARMY), Army of the Republic of Vietnam (ARVN), andRepublic of Vietnam Air Force (RVNAF) base located in Soc Trang in southern Vietnam. Due to its location and isolation, it had to be self-supporting. An aerial view of the Soc Trang Army Airfield during the monsoon season with flooded rice paddies is provided in Figure 33.

Figure 33. Aerial View of Soc Trang Airfield in Monsoon Season

Unlike the Can Tho Army Airfield located near the Binh Thuy Air Base and the significant city, Can Tho, the Soc Trang Army Airfield was isolated and needed to be self-supporting.

Figure 34. Soc Trang Airfield with Key 13th CAB and Support Elements, 1970

A 1970 picture illustrating critical elements of the 13th CAB and support elements is provided in Figure 34. The 13th CAB Headquarters, which included the Commander and Executive Officer Offices, the S-3 Operations Office, and Tactical Operations Center (TOC), and quarters are identified along with the 121st AHC "Soc Trang Tigers" and 336th AHC Warriors Company areas, hangars, and hooches. Support elements identified include the mess hall, post chapel, post theater, officers' club (Tiger Den), control tower, ammo dump, and even a swimming pool.

The major 13th CAB combat units at Soc Trang AAF were the 121st AHC (Tigers and Vikings) and the 336th AHC (Warriors and Thunder Birds). Also, C Troop, 16th Air Cav (Darkhorse) in much of 1970, worked out of the Soc Trang AAF. A summary of each of these units follows:

THE 121ST AHC SOC TRANG TIGERS

On 15 December 1961, the 93d Transportation Company (Helicopter) departed the continental U.S. for Vietnam to provide aerial transportation for the armed forces of the Republic of Vietnam (RVN). First stationed at Da Nang, it gained an outstanding reputation as it helped develop heliborne assault techniques, supply methods, and medical evacuation operations.

The unit then relocated to Soc Trang, RVN, in September of 1962 and became the first Army helicopter company in the Mekong Delta. Because of its isolated location, the commander of the 93d Transportation Company was also the post commander of Soc Trang AAF.

As such, the unit was responsible for all aspects of its members' lives, e.g., medical, security, air traffic control, communications, and much more. But these responsibilities in no way slowed the pace of operations. In December 1962, the unit logged 1,000 hours of flying time.

Figure 35A. Soc Trang Welcome Sign w/ "Tuffy."

In January 1963, the unit obtained a fifteen-week-old Bengal tiger. Although the tiger, named "Tuffy," only stayed with the unit until June of 1963, his legacy was to the name "the Soc Trang Tigers." 121st AHC consisted of the Soc Trang Tigers, with the gunship platoon being the Vikings. Illustrated in Figure 35A is a picture of "Tuffy" and handlers under the welcome sign.

Figure 35B is an updated welcome sign from "The Famed Soc Trang Tigers," with an "Elevation of 2 Feet." "Many accounts of dedication, skill, and heroism occurred during the 121st Aviation Company served at Soc Trang (1962 to 1970). The gunship platoon was the Vikings. A grateful nation gave many awards, both unit, and individual, in recognition of the contributions of this outstanding unit and the men who served her. Finally, in October 1970, the 121st AHC stopped reporting to the 13th CAB operationally. Instead, they started receiving missions from the local

Figure 35B. Later Soc Trang 12th Welcome Sign

Vietnamese National Air Force (VNAF) Wing. Their primary mission was to train Vietnamese officers and enlisted men in the details of the transport side of the 121st AHC's mission (gunships were not transferred to the VNAF). In December 1970, the 121st AHC ceased operations with the transfer complete.

336TH ASSAULT HELICOPTER COMPANY - "WARRIORS" AND "THUNDERBIRDS" SOC TRANG

The 336th Assault Helicopter Company (AHC) deployed to the Republic of Viet Nam as "A" Company, 101st Aviation Battalion, and was based at Soc Trang. In September 1966, the 101st Airborne Division relocated north to the I CTZ. "A" Company was also relocated north to I Corps and became known as the 336th AHC. The 336th AHC was relocated back to Soc Trang and operated there until October 1970, when the Company stood down with its aircraft transferred to the VNAF. The last Aviation Unit to leave Soc Trang. The 336th AHC Welcome Sign and Shield are illustrated in Figure 36.

Figure 36. The 336th AHC Welcome Sign and Warriors Shield

C TROOP, 16ᵀᴴ CAV EVOLUTION TO THE MEKONG DELTA

The C Troop, 16ᵗʰ Cav evolved from the 1ˢᵗ Aviation Battalion, 1ˢᵗ Infantry Division, "Big Red One," in Phu Loi, Vietnam, 1969. When the 1ˢᵗ Infantry Division deployed back to the USA, most elements of the 1st Infantry rotated back to "the world."

Those with time remaining in D Troop, Companies A & B, 1st Aviation Battalion, and a few others were combined to form C Troop (AIR) 16th Cavalry (Darkhorse) and assigned to the 1st Aviation Brigade in the Mekong Delta at Can Tho and then Soc Trang AAFs in Spring 1970. This evolution is in Figure 37.

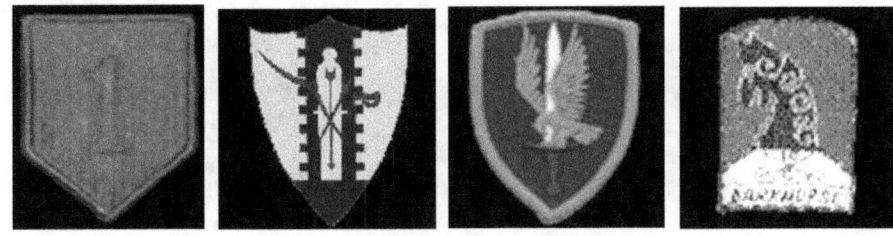

Figure 37. Symbolic Evolution Big Red One to C Troop, 16th Cav

In an Air Cavalry Troop, there was a slick platoon of 6-8 UH-1D/H aircraft. There was also a gun platoon of 8-9 UH-1 series gunships or attack helicopter Cobras, similar to the smooth and gun platoons of an AHC. However, there was also a "scout" platoon of 8-9 Light Observation Helicopters (commonly called "loaches"). Initially, these were Bell OH-13s or Hiller OH-23s and then OH-6A Cayuses. The C Troop, 16th Cav, had UH-1D lift ships, AH-1G Cobra attack helicopters, and OH-6A Cayuses, as illustrated in Figure 38. The Patches for the Attack Helicopter *Mustangs*, the Lift Ship *Four Horseman*, and the Scout "Loach" *Outcasts* are shown in Figure 39.

The C Troop, 16th Cav's new Area of Operations (AO), was the Mekong Delta, with the 13th CAB at the Soc Trang AAF to introduce "Charlie" to Darkhorse. No longer supporting American units, they began direct support to the 21st ARVN Division as part of the 13th CAB Task Force Guardian Mission. Darkhorse also did occasional work with the Navy SEAL Teams in the area. In May 1970, they were flying regular missions in Cambodia. They had quickly become a favorite of the 21st ARVN

Figure 38. C Troop, 16th Cav Aircraft, UH-1D, OH-6A, and AH-1G Helicopters

*Figure 39. The Patches for the Attack Helicopter Mustangs; the Lift
Ship Four Horseman, and the Scout "Loach" Outcasts*

Division, along with the 121st AHC Soc Trang Tigers. The 21st ARVN Division Operations Center often requested Darkhorse and clearly stated, "No VNAF." Once again, this illustrated the animosity between ARVN and VNAF. In August 1970, Darkhorse was on the move again. This time up to Can Tho AAF, as a result of Helicopter Vietnamization, Soc Trang AAF was being turned over to the VNAF and their U.S. Air Force Advisors on November 4th, 1970. The first turnover of an AAF.

The 1968 Tet offensive and how the 121st and 336th AHCs Gunships and Lift Ships saved the day for Soc Trang and Vinh Long and kept control of the Mekong Delta

INTRODUCTION TO 1968 TET OFFENSIVE AND 13TH CAB DEFENSE OF SOC TRANG/CAN THO/VINH LONG ARMY AVIATION AIRFIELDS (AAAFS)

At 3 a.m. on Jan. 31, 1968, North Vietnamese and Vietcong forces launched a wave of simultaneous attacks on South Vietnamese and American forces in major cities, towns, and military bases throughout South Vietnam, as illustrated in Figure 40.

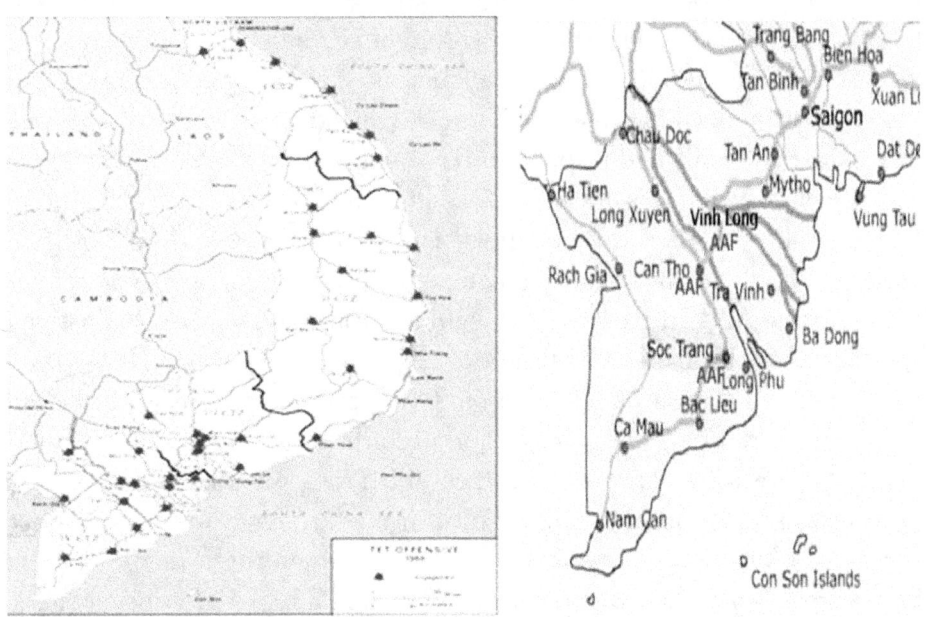

Figure 40. VC and NVA Tet '68 Attacks.   Figure 41. Key AAF Attacks

The fighting, the heaviest and most sustained of the Vietnam War, coincided with the Lunar New Year, or Tet, and it has been called the Tet '68 Offensive ever since. In the Mekong Delta, the principal Army Airfields of Soc Trang, Can Tho, and Vinh Long (Figure 41), were immediately recognized as prime targets for heavy attacks by Viet Cong infantry and artillery units and sapper teams.

All three airfields were under significant attack with the threat of being overrun by Viet Cong and NVA forces. The key to keeping this from happening was the leadership of key individuals and the rapid repositioning of ARVN and Ranger Forces, possible through airmobile operations. The Commanding Officers for the 121st AHC and 336th AHC, using their gunship platoons, Vikings and Thunderbirds, kept the Soc Trang and Vinh Long AAF and their adjoining cities from being overrun by the Viet Cong and NVA forces. This kept the Mekong Delta from VC and NVA control. These military victories could have been followed by complete control of the Mekong Delta, the population and agricultural center of South Vietnam.

Detailed Description of the Defense of Soc Trang City and AAF with Gunships (13th CAB Tet '68 Report)

On 31 January 1968, the Viet Cong initiated a ground attack against the town of Soc Trang while simultaneously conducting a mortar and recoilless rifle attack against Soc Trang Army Airfield. At 0330, the Viet Cong shelled the airfield using 75mm recoilless rifles and 82mm mortars. The Copperheads and Thunderbirds gunships immediately responded by scrambling to the area of known enemy positions and taking them under attack. Utilizing organic gunships from both helicopter companies at Soc Trang AAF, the command elements managed to have at least one light fire team airborne at all times.

A second mortar attack began at 0405, and it soon became evident that this mortar barrage was just a prelude to a more concentrated attack. Sporadic fire on the outskirt of Soc Trang was the first indication of a Viet Cong ground assault. The downtown area soon came under attack by a large Viet Cong force estimated to be a reinforced battalion. It concentrated its attack on government and U.S. Army installations. By 0630, the fighting in downtown Soc Trang was very intense, with the Viet Cong gaining momentum and an initiative resulting in the gunships (UH-1Bs) being called upon to assist the ARVN troops. Although the aircraft received intense automatic weapons and small arms fire, they continued to make target attacks throughout the day. At first, the gunships concentrated their attacks on the southwest corner of town where the Viet Cong were making concentrated efforts to overrun the Administrative and Logistics Compound—using B40 rockets, 75mm recoilless rifles, and large numbers of autonomic weapons.

The Viet Cong slowly pushed the ARVN troops out of the compound. As a result, a nine-man detachment from the 78th Artillery was trapped in the compound, but due to their valiant defenses, they were able to slow the Viet Cong Advance. The gunships made continuous attacks on the advancing forces. The superb fire support rendered allowed the nine-man detachment to escape and move into a more secure position where they could defend against the rapid-moving forces.

While providing the compound with this vital support, the gunships made severe firing passes on enemy positions strategically located along the Air and Logistics (A&L) Compound routes. Their devastating firepower and accurate concentrations soon forced the Viet Cong to move away from their blocking positions allowing the ARVN troops to advance to positions where they could more readily assist the defenders of the A&L Compound.

The gunships flew all day, and at dusk, they began the necessary preparations for night operations. One fireteam remained over the airfield in a defensive role while a second fire team assisted the ARVN troops by giving them fire support cover over most of the town. UH-1D flare ships remained airborne at all times providing sufficient light over the airfield and town. Mortar attacks were initiated against the airfield at 0215, and the gunships used their superior firepower to quell the attack.

At dawn, on 1 February, the gunships began an all-out support operation of ARVN troops who were starting to take the offensive in the downtown area of Soc Trang. The Mission Commander, Major Carl H. McNair Jr., 121st AHC Commander, was the coordinating control for all supporting air operations.

The transports were dispatched for airlift elements of the 31st Infantry Regiment from Vi Thanh to Soc Trang, where they were to reinforce troops of the 33rd Regiment. CH-47 helicopters also assisted by moving elements of the 44th Rangers from Vi Thanh to Soc Trang. The gunships combined their efforts with ground forces to dislodge the Viet Cong from their defensive positions and slowly began to push them out of town. The gunships kept the Viet Cong moving, never allowing them a chance to regroup to take cover and organize good defensive positions.

As the Viet Cong moved into a defensive position on the southeast edge of town, airstrike after airstrike was used to destroy their positions. As the enemy moved into a large pagoda in the north-central part of the city, the gunships put a concentrated barrage on the area around the pagoda. Despite heavy automatic weapons fire, the armed helicopters continued their assault until the VC-infested area began to burn uncontrollably. The Viet Cong were forced to evacuate the buildings and ran into the streets and open rice fields north of the city, thus allowing the friendly forces to engage in hot pursuit. As the friendly forces began their concentrated assault on the Viet Cong, the armed helicopters set up a "daisy chain" over the retreating enemy and kept the VC completely disrupted. Through the combined efforts of gunships and ARVN units, the Viet Cong units were routed and sustained heavy casualties.

On 2 February, the transports departed for Can Tho while the gunships continued to aid in the Soc Trang ARVN offensive against the Viet Cong forces. The last pockets of resistance were cleaned up in the city, and the ARVN forces began the final pursuit of the Viet Cong along the rivers and canals. Gunships continuously broke up Viet Cong's hasty defensive positions allowing the ARVN troops to continue their pursuit. With the town secured, the gunships moved to the southeast, where the Viet Cong were reported to be digging in. Finding the enemy in a large pagoda on the outskirts of My Xuyen, the gunships repeated their performance, striking the area around the pagoda. As the VC evacuated and ran, the gunships used their devastating firepower and followed in hot pursuit routing his remaining forces. Intelligence documents revealed that this unit had been the headquarters group for the Soc Trang Mobile Battalion. The destruction of this battalion was instrumental in preventing any other VC significant offensives against the city and the airfield. Operations continued around the city and airfield for the next three days as remaining Viet Cong forces were pushed out of the area.

The timely support of the gunships enabled the friendly forces to disrupt the Viet Cong offensive. The devastating firepower of the armed helicopters was responsible for the successful completion of the Soc Trang operation and enabled the friendly elements to secure the airfield and the city. Without the support of helicopters, both armed and transport, the friendly forces might not have been able to secure the area, and it is possible that the attacking Viet Cong forces could have overrun the city and airfield.

As attested to by the record of performance and the results attained, Army Aviation assets contributed to the successful defense of major installations. The support rendered was not available from other sources. Because of the flexibility, responsiveness, and perseverance of the airmobile units, losses of both U. S. and Vietnamese personnel and property were minimized. What started as another Viet Cong offensive maneuver to capture the city of Soc Trang resulted in a significant Viet Cong defeat.

Because of the close-in fighting and the nature of the operations in and around Soc Trang, defensive and offensive operations became almost synonymous. Army Aviation assets were effectively utilized offensively in that they succeeded in disrupting enemy units moving for new attacks. It was the immediate reaction capability of armed helicopters. UH-1D flare ships and lightships actively engaged the enemy during the crucial hours of darkness throughout the "Tet Offensive."

DETAILED DESCRIPTION OF DEFENSE AND NIGHT ASSAULT ON VINH LONG AAF (13TH CAB TET '68 REPORT)

Throughout the "Tet '68" offensive, Army Aviation assets were noted as being at the forefront of offensive and defensive operations directed towards blunting VC attacks and defeating their attempts to overwhelm allied forces and capture installations. In the Mekong Delta, the principal Army Airfields of Soc Trang, Can Tho, and Vinh Long were immediately recognized as prime targets for heavy attacks by Viet Cong infantry and artillery units and sapper teams. After the period of the ceasefire had been announced and only thirty-six hours after it had been officially retracted, the 121st Assault Helicopter Company and attached units were scrambled by the Army Aviation Element, IV Corps. It was announced that Vinh Long Army Airfield was in a critical situation and in danger of being overrun. The Tigers were called upon to assist their sister assault helicopter companies by carrying out a night combat assault into Vinh Long despite intense mortar, recoilless rifle, small arms, and automatic weapons fire.

The 121st AHC was selected as the primary unit for the airlift, with Major Carl H. McNair Jr. designated Mission Commander. Major McNair immediately flew to Can Tho for the detailed briefing conducted at HQ, 13th Combat Aviation Battalion. In the meantime, the Tiger crews at Soc Trang were alerted, and Lt. William E. Hattaway was given a short mission briefing. Although integrated into the perimeter defense, the aircraft crew prepared their aircraft and stood by for a scramble mission. Upon receiving orders to get airborne, the lead aircraft started its engine, scrambling all other crewmembers immediately to their assigned aircraft. At 0015, Soc Trang itself was hit by a 75MM R. R. attack. However, the crews operated their ships for the scramble, and the lead cranked his engine at 0020, and the entire flight was airborne by 0030. The flight immediately proceeded to Can Tho, where they refueled and picked up any U.S. personnel available as troops. While at Can Tho, a flight of UH-1Ds from the 336th Assault Helicopter Company joined the Tigers, giving the Mission Commander a combined total of fourteen (14) UH-1D aircraft. The flight was shut down, and Major McNair briefed the Aircraft Commanders on the mission, emphasizing the proposed approach and emergency procedures. Because of the intense enemy concentrations around Vinh Long and some VC penetrations had been made on the south side of the runway, it was recommended that the aircraft approach from Northeast to Southeast from over the river and use only the last one-fourth of runway 08 for landing.

A flare ship was prepared to light up the runway if needed for the landing since it was decided that the transports would not use their landing lights unless required to prevent an accident. The Aircraft Commanders then briefed their crews and instructed the troops that upon arrival at Vinh Long, they were to immediately exit the aircraft and move into the revetments on the south side of the runway. At 0130, the flight was once again airborne and enroute to Vinh Long. Two light fire teams of AH-1G attack helicopters escorted the flight while a light fire team of UH-1C gunships orbited over Vinh Long. The troop transports were told not to suppress under any circumstances because of the proximity of friendly forces.

As the flight approached Vinh Long, the Vs of five formations were changed to the trail, and the flight commenced its approach. The gunships began to suppress the area surrounding the airstrip, including the area just off the approach end of runway 080, causing a secondary explosion. Simultaneously, two VC automatic weapons began to fire at the flight from a position approximately 75 meters to the right of the runway. On a short final, the flight started to receive heavy ground fire from several other locations on the same side of the runway. The transports continued their approach disregarding the apparent danger, and successfully inserted the troops. As each unloaded, the pilot immediately initiated his take-off so that a minimum amount of time was spent on the ground, significantly reducing the possibility of being mortared. Departure instructions were that the flight was to make a left turn before reaching the end of the runway. As the aircraft gained airspeed and altitude, the Viet Cong opened up from both sides of the runway. Gunships immediately used up their remaining ordnance to suppress the intense fire being received. The transports gained altitude, formed in Vs of five, and proceeded to Soc Trang. Miraculously, no aircraft was hit by the intense fire, although numerous tracers were seen passing between the skids and the aircraft's fuselage while on final approach and take-off. The excellent cover and suppression provided by the gunships combined with the insertion speed, enabled the lift to be carried out without sustaining casualties and/or damage to aircraft. The timely insertion allowed the defenders at Vinh Long to reconstitute their perimeter defense, secure the airfield, and begin repairs on damaged aircraft.

Major McNair's contributions to saving both Soc Trang and Vinh Long were outstanding and are summarized below:

- Served as Commander of the 121st AHC, "Soc Trang Tigers," and served as the Mission Commander for the defense of the Soc Tran AAF on January 31 and February 1, 1968, during the 1968 Tet Offensive

- He was then called by the 13th CAB Commander in Can Tho to be the Mission Commander to lead the Night Air Assault on the Vinh Long AAF after the Airfield Commander was killed and to keep it from being overrun by the VC/NVA on February 2, 1968

He logged a record 3,840 combat hours in one company for one month, Jan-Feb 1968. Later, in South Vietnam, he served as the G-3 of the 164th Combat Aviation Group and later as Commander of the 145th Combat Aviation Battalion. Major Carl H. McNair went on to become a Major General and the first Director of Army Aviation at Fort Rucker, AL, in 1983 when Army Aviation became a Combined Arms Branch, Figure 42A. He was also a Georgia Tech distinguished graduate and congratulated Dr. Schrage on Georgia Tech's Vertical Lift Research Center of Excellence (VLRCOE) success, Figure 42B.

Figure 42A. MG Carl H. McNair.

Figure 42B. MG(Ret) McNair with Dr. Schrage

COMBAT ASSESSMENT OF THE 307TH CAB ARMED HELICOPTER SUPPORT FOLLOWING TET '68

Following the successful use of armed helicopters in saving the three major AAFs in the Mekong Delta, i.e., Can Tho, Soc Trang, and Vinh Long, during the Tet '68 VC and VNA attacks, the Commanding Officer of the 164th CAG, COL Robert L. McDaniel, sought feedback from his major customers, e.g., Major General Nguyen Van Minh, Commanding General of the 21st ARVN Division and 42nd Defense Tactical Area (DTA), COL Josiah A. Wallace, JR, Province Senior Advisor, MACV TEAM 56, Phong Dinh Province, and COL William M. Calnan, Senior Advisor, 42nd DTA. They also provided input to a questionnaire on Combat Assessment of Armed Helicopters. (Documented in the 307th CAB Army Aviation Operations Report, {29 January through 29 February 1968} TET, 17, April 1968)

A Summary of the 307th CAB and the Major Customer Responses to each item of the Armed Helicopter Questionnaire:

1. What type of missions have your gunships performed since 31 Dec 67?

    a. Combat-in-cities and airmobile operations)

    a. Direct fire support (Landing Zone {LZ} preparation)

    b. Visual reconnaissance

    c. Armed reconnaissance

    d. Escort missions

    e. Screening force for ARVN troops

    f. Advanced guard for assaulting ARVN troops

    g. Conduct of Items a through f at night a. Close air support (Combat-in-cities and airmobile operations)

    h. Direct fire support (LZ preparation)

    i. Visual reconnaissance

    j. Armed reconnaissance

    k. Escort missions

    l. Screening force for ARVN troops

    m. Advanced guard for assaulting ARVN troops

    n. Conduct of Items a through f at night

2. Could these missions have been accomplished by other means? If not, why?

    It is highly doubtful that anything other than armed helicopters could have performed these missions. The gunships were immediately available and responsive to the advisors in IV Corps. Requests followed a much more precise command line for the contact of armed helicopters than other assets. The missions assigned were almost entirely short-fused, and time was the prime factor. Air Force tactical air would not engage targets close to the troops without first asking for relief of all responsibility for the safety of the ground elements. A great majority of all attacks started during the hours of darkness; the armed helicopter gunships continued to operate with artificial illumination and provided identical support to that possible during daylight hours. No other airborne weapons platform was available in most of the attacks. During many attacks, it was found that too much reliance was placed upon the gunships when artillery was available. It was challenging to get ARVN artillery to fill the gaps when gunships had to engage the target, having seen muzzle flashes of incoming weapons. The artillery at Soc Trang did not even have prior concentrations to employ proper shift techniques. Therefore, most counter-battery fire was affected by the helicopter gunships.

3. During the Tet Offensive, did your gunships play an essential role? The gunships of this battalion were most probably the deciding factor in the defeat of the Viet Cong in Can Tho city and Can Tho airfield. The primary principle, the quick massing of firepower, is possible with gunships, making the gunship an asset. The 235th AHC destroyed 1,690 structures, 430 sampans, killed 894 Viet Cong, and wounded 653 during the 29 January - 29 February time period. The majority of the claims are for the cities of Can Tho, Tra Vinh, My Tho, Ben Tre, Vinh Long, Moc Hoa, and Tieu Can. The greatest asset responsible for turning the tide of the VC offensive in IV Corps was the armed helicopters of the 164th CAG.

   The documentation on the effectiveness of the 121st AHC and 336th AHC gunships in saving Soc Trang and Soc Trang AAF and Vinh Long and Vinh Long AAF is in the Section D descriptions in this chapter and were not included in the 307th CAB Army Aviation Operations (29 January through 29 February 1968) Tet.

4. During the Tet Offensive, did Air Force aircraft share in the defense of your airfield, and to what extent?
   Can Tho Airfield: The first Air Force strike was at the first light of the morning following the night of the first attack on 31 January 1968. As mentioned, armed helicopters were airborne within 3 minutes of the attack. On 31 January and for the next five days, daylight strikes were employed to within 3,000 meters of the airfield. Air Force "Spooky"(AC-47) aircraft were generally available, but a standoff altitude of 3,000 feet does not allow for effective observation of the engaged target. The cloud layer on 31 January 1968 precluded the use of "Spooky."

5. What increase in capabilities over current gunships are desirable in future gunships? Why?

    a. A larger caliber point fire weapon, preferably 20 or 30mm cannon on a 360-degree flexible mount with a greater selectivity of warheads, is needed. Weapons of this type would allow a standoff capability to engage large-caliber anti-aircraft weapons such as 50 cal. or 12.7mm before entering an area to destroy personnel and positions.

    b. The ability to carry a larger ordnance load is desirable. This capability would allow for a greater selectivity with all weapons while in flight and longer station time.

    c. More airspeed in straight and level flight is desirable. The AH-1G has adequate speed in a diving attack and for visual reconnaissance low-level. However, to disengage from a low-level target attack or a target of opportunity engaged low-level, more dash speed is necessary to ensure less vulnerability.

6. Please give all factual written statistical data available, based on the experiences of your unit, which would support the following contentions:

    a. The UH-1 gunship is providing vital support not available from the Air Force or any other means.

    The UH-1 gunship provides all the support mentioned in question number one. It would be impossible to escort an air column of slicks with any fixed-wing attack aircraft we presently have. Weather restricts fixed-wing aircraft that does not affect helicopters. The maneuverability of a helicopter air column dictates that it be escorted with something of comparable ability. The UH-1 gunships can spot and deliver exact firepower upon the enemy with immediate response. The capability to recon by fire and see the results can only be done by gunships.

    b. The AH-1G can provide better support than the UH-1 gunships.

       The AH-1G (Cobra) has been highly successful in combat operations here in IV Corps. This aircraft has been able to assume all the missions presently assigned to older UH-1B/C model gunships. The speed and the increased ordnance load coupled with the versatility and accuracy of the armament systems have made the aircraft exceed our gunships' capabilities and give us the ability to provide broader tactical capabilities.

7. If gunships had not been available in your area of operation during the Tet Offensive, what would the probable results have been? Why? The consensus is that the armed helicopters in IV Corps were the most singular factor in turning back the Viet Cong offensive. The VC would most probably have occupied all three of the major airfields for a while. The length of the offensive would have been much longer. The gunships were the only direct fire support for the airfields during some phases of the initial offensive. The tenure of the Viet Cong in cities was always shortened by the arrival of gunships with their ability to ferret out the VC in their hiding places in the cities and catch the VC in the open as he attempted to retreat.

This summary of the effectiveness of Army Aviation's attack helicopters, gunships, and lift ships in South Vietnam, especially during the '68 Tet Offensive, became extremely valuable in my next career with Aviation Systems Command (AVSCOM). This included developing the Next Generation of Army Aviation Systems, as described in the second book of this Trilogy. Within three years after my return from South Vietnam, after I attended the Field Artillery Officers Advanced Course and received a Master's Degree in Aerospace Engineering from Georgia Tech, I was assigned to the AVSCOM. As an aerospace engineer, I evaluated the Army Advanced Attack Helicopter (AAH) and Utility Tactical Transport Aircraft System (UTTAS) candidates against the evolving operational requirements for new attack and utility helicopters to replace the AH-1 Cobras and UH-1 Hueys. This included serving as Aeroelasticity Dynamics and Vibration (ADV) evaluation engineer on the UTTAS and AAH prototypes and their Source Selection Evaluation Boards (SSEBs). Vibration and Dynamics were critical issues, and my role in their evaluation will be discussed in book two of this Trilogy.

EXPERIENCES AS ASST S-3/S-3 IN 13TH CAB AND MY STORIES JULY—NOVEMBER 1970

My arrival at Soc Trang AAF as the Assistant S-3 for the 13th CAB in July 1970 gave me a unique opportunity to build on my air mobile experiences with the 162nd AHC, as well as to use my experiences as a battery commander of an Honest John Missile Battery in Germany in 1968-1969.

In both experiences, I served as a junior officer in a higher rank position, e.g., in Germany, the Battery Commander position was at the Captain level, while I was a 1LT. While I was assigned as the Assistant S-3, a Captain position, in the 13th CAB, I was soon given the authority and responsibility of the S-3, which was a senior Major position.

Due to the tour rotation of the then-current S-3, I had experience leading airmobile operations and knowledge of the Mekong Delta. LTC Robert Sauers, the 13th CAB Commander, recognized these capabilities and gave me the authority to run the day-to-day operations out of the 13th CAB S-3 TOC and seek new opportunities for the 13th CAB.

My quarters, office, and surroundings at Soc Trang AAF were superior to the 162nd AHC tent city. Shown in Figure 43A is the Entrance to the 13th CAB HQS; Figure 43B is CPT Schrage leaving the S-3 TOC. Figure 43C is CPT Schrage outside the 13th CAB S-3 Shop entrance.

Figure 43C. CPT Schrage outside the 13th CAB S-3 Shop Entrance

Figure 43A is the Entrance to the 13th CAB HQS

Figure 43B CPT Schrage leaving the S-3 TOC.

Figure 44A. CPT Schrage's Jeep Figure

Figure 44B. CPT Schrage's Helicopter.   *Figure 44C.13thCAB Shield*

Figure 45. Personnel in the Guardian 13th CAB S-3 Shop, 1970

In Figure 45 seated at the center of the above picture is MAJ Burnum Melton, S-3 and on his left is CPT Dan Schrage, Asst S-3/ S-3. Other personnel included NCOs, enlisted personnel, and warrant officers.

A truly remarkable group of dedicated individuals who worked in the TOC 24 hours a day in shifts to oversee and help run operations throughout the Mekong Delta. When MAJ Melton came to the 13th CAB in August 1970, CPT Schrage ran the Guardian S-3 Shop with the approval of LTC Robert Sauers, the 13th CAB Commander. MAJ Melton was an excellent officer but had been a fixed-wing aircraft pilot who transitioned to rotary-wing en route to Vietnam with no airmobile operations experience. Therefore, I continued to run the S-3 TOC and conducted Air Mission Commander operations with the Navy SEALs and 13th CAB units. LTC Sauers said he would give MAJ Melton command of the 221st Reconnaissance Airplane Company in Vinh Long when it became available. He stated he would not request a new S-3. In my final South Vietnam Officer Efficiency Report (OER), July 1970-January 1971, both MAJ Burnum E. Melton, the OER Rater, and MAJ Charles A. Lepore, 13th CAB Executive Officer and the OER Endorser, stated that CPT Schrage was *"the most outstanding officer they had known."* This helped support my "Why" questions in *"As Usual Guardian was Perfect in All Respects."*

THE FOLLOWING STORIES REFLECT THE THE CHALLENGE AND REWARDING EXPERIENCES I HAD AS S-3 13TH CAB

1. **Coordination of Airmobile Operations in the Mekong Delta**

    The Guardian AO included most of the Mekong Delta. I got to work closely with all aviation elements organic to the 13th CAB and those called in support of primary operations, especially where major Viet Cong/NVA encounters occurred throughout the Mekong Delta. This involved working with the 13th CAB AHCs, Air Cav Troop, VNAF commanders, and 21st ARVN Division commanders and advisors. The U.S. Advisors to the ARVN Divisions and Regional and Popular Forces and the Army Special Operations commanders along the Cambodian Border had very challenging jobs. To give them some breaks, I often sent helicopter lift ships to pick them up and bring them to Can Tho or Soc Trang for a few days of relief.

2. **Support for Navy SEALs POW Operations to Free U.S. POWs in the Mekong Delta**

    It was known in 1970 that at least 12 U.S. prisoners were being held by the Viet Cong/NVA in Mekong Delta POW camps. As discussed in Chapter two, Section 1.0, during the Vietnam War, the U Minh Forest was an area of thick mangrove swamps located in the northwest quadrant of the Ca Mau Peninsula. It was a feared VC, and NVA stronghold and the site of more than forty confirmed communist prisons and detention facilities.

    In the 162nd AHC support operations for the Navy SEALs in August 1970, Bill Tuttle, a 162nd AHC AMC, was flying support missions for the Navy SEALs out of the SEA FLOAT in the Can Mau Peninsula. The third time out, the Navy Sea Wolves gunships escorted his ship, so the SEALs were interested in that area, too. It turns out the SEALs were trying to pinpoint the location of a base camp/POW compound. So, they pulled a six-ship, no-prep raid in that area—the six ships landed and stayed in place instead of pulling pitch and getting out of there.

The local bad guys must have bugged out while en route because you could still smell the faint odor of wood smoke. The ARVN grunts found the compound underground—three stories, complete with a generator room, pumps, a decent-sized hospital, and a clothing and equipment manufacturing area with fifty Singer sewing machines, but no live POWs—just dog tags stuck in the walls. The VC had kept the POWs in 3'x3'x5' cages dug into the sides of the tunnels, and when they died, they just collapsed each cage roof to bury them. They tamped the dirt down and tucked their dog tag chains into the wall. Bill Tuttle stated that it definitely reinforced your desire to go out fighting if you went down while flying a single ship and became a POW.

The official reason for the presence in Vietnam of the commandos called SEALs—an acronym for Sea, Air, Land— had been "intelligence collection." But according to sources closely connected to the program, the most important reason for their remaining in South Vietnam over their last years, 1970-71, was their role as a contingency force for rescuing American POWs. While I was the 13th CAB S-3, I was contacted by a SEAL Team Platoon leader about using our aviation assets to insert Navy SEAL Teams to try and free U.S. POWs in the Mekong Delta.

Only five Americans were rescued from captivity during the entire Vietnam War. One died after rescue from wounds inflicted by his guards before they ran from the rescue forces. It was the only intended rescue. The others were unplanned rescues. One took place while the prisoner was being escorted from his capture to a prison camp, and another involved a helicopter assault in an area that turned out to hold prisoners. An American POW broke loose and ran to the helicopters.

The Navy Seal's Strategy in the Mekong Delta strategy was to get intelligence from The Chiêu Hồi Program (also spelled "chu hoi" or "chu-hoi" in English), loosely translated as "Open Arms"). It was an initiative by the **South Vietnamese** to encourage **defection** by the **Viet Cong**(V.C.) and their supporters to the Government's side during the **Vietnam War**. There were 101,511 Viet Cong who defected under the program. Chieu Hoi were Viet Cong who were captured or turned themselves in. Overall, the Chieu Hoi program was considered successful. Those who surrendered were known as "Hoi Chanh" and were often integrated into allied units as Kit Carson Scouts, operating in the same area where they had defected. Many made significant contributions to the effectiveness of U.S. units and often distinguished themselves, earning decorations as high as the Silver Star. The program was relatively inexpensive and removed over 100,000 combatants from the field (assuming the accuracy of the numbers recorded and the sincerity of the defections).

Several 13th CAB operations with the Navy SEAL Platoon operated out of Binh Thuy AFB north of Can Tho. The first 13th CAB operation with the Navy SEAL Team was based on a Hoi Chanh who had worked as a Viet Cong guard at a VC POW Camp in a mangrove-covered area somewhat south of Can Tho. It was believed to include two U.S. prisoners, along with ARVN POWs. The 13th CAB "package" supporting this operation included two UH-1D/H lift ships carrying the SEAL Team with me flying the 13th CAB S-3 OH-58A as the AMC with the Navy SEAL Platoon Leader. It included having two 13th CAB UH-1C gunships loitering behind for call-in if necessary. The planned operation kicked off with an assault landing next to the VC POW Camp Area at first light.

As the two lift ships approached and were landing in the VC POW Camp Area, the first lift ship AC signaled me by radio that the Chieu Hoi in his aircraft was waving his hands not to land there but to proceed about another 500 meters further to a different landing area. Since he was our intelligence, the Navy SEAL Platoon Leader and I instructed the ACs of both aircraft to proceed further. Once the SEAL Team landed, they moved into the mangrove forest where a VC POW Camp, designated #2 Camp, was located. Shortly after the Navy SEAL Team departed the aircraft and approached the VC POW Camp #2, an intense firefight broke out. I called in the UH-1C gunships to support the Navy SEAL Team. After about 30 minutes of fighting, the Navy SEAL Team progressed toward the first VC POW Camp #1, about 50 yards from the first landing. They found a small hooch that illustrated that a complete upheaval had happened inside. It also included two sets of chains, typical of the type used on U.S. POWs. It was concluded that panic had set in this first VC POW Camp when the two aircraft carrying the Navy Seals Team had first landed. The aircraft repositioning to the second landing, which the Chieu Hoi had encouraged, had allowed the VC to escape with the two U.S. POWs. While some ARVN POWs were released, it was found out later, in the Navy Seals Team debriefing, that the Chieu Hoi had been a guard at the second camp and wanted to have the VC guards at this camp killed, so his identity was not recognized by the VC guards. Therefore, an opportunity to release two U.S. POWs was narrowly missed.

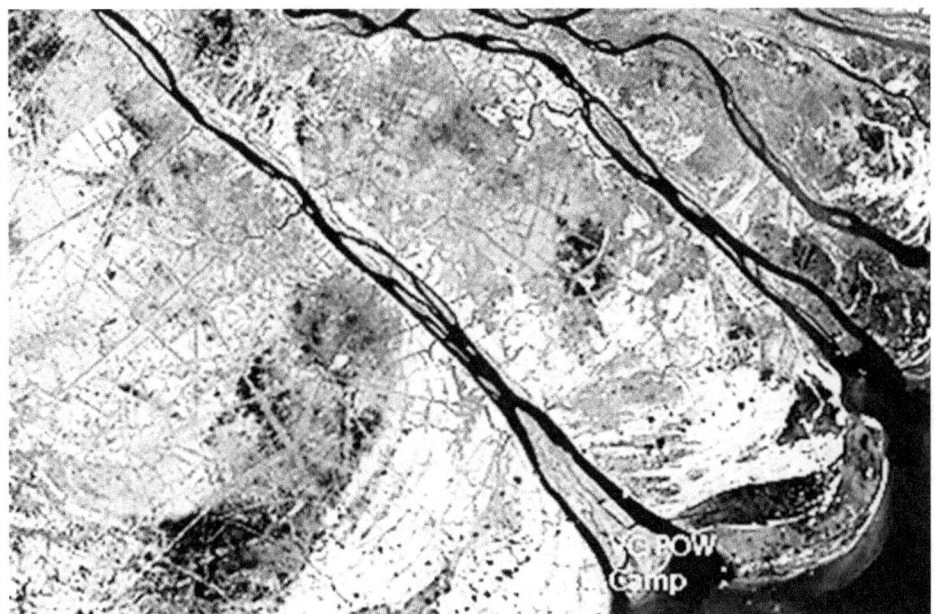

Figure 46A. Identified VC POW Camp

The last 13th CAB operation with the Navy SEAL Team was based on the intelligence that a VC POW Camp with U.S. POWs was in the heavy mangrove swamp on the island where the Bassac River flowed into the South China Sea, as illustrated in Figure 46A. The same aircraft package of two UH-1D/H lift ships, a Command & Control Ship of the 13th CAB S-3 OH-58A Kiowa with myself as the AMC assisted by the Navy SEAL Platoon Leader and with two UH-1C gunships on standby.

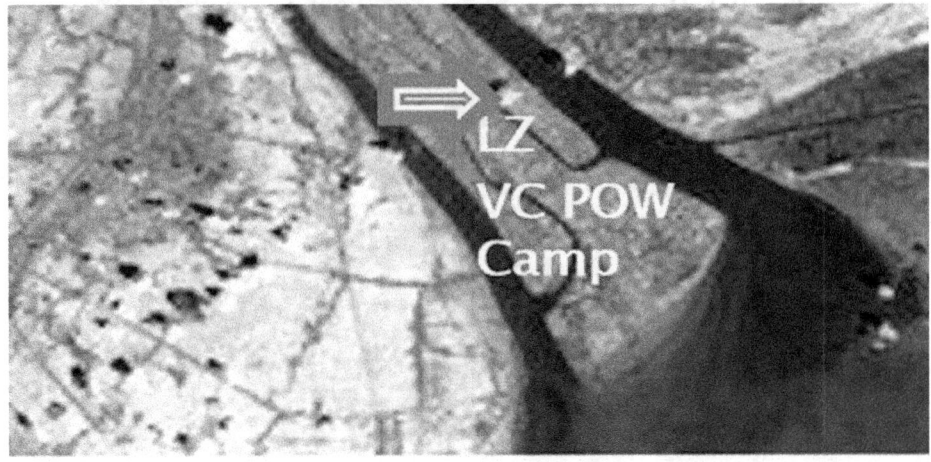

Figure 46B. Planned Landing Zone

Figure 46B illustrates the flight path arrow into the only practical Landing Zone on the Island. The Navy SEAL Team was successfully inserted into the Landing Zone. Shortly after they were inserted, an intensive firefight with the VC took place, including mortars and exploding booby traps that looked like a fourth of July demonstration. It wasn't long afterward that we received a frantic call from the Navy SEAL Team Leader that he had wounded SEALs and needed to be extracted. This led to me vectoring one of the UH-1D/H lift ships along the original approach path into the LZ. Upon final approach, the UH-1D/H AC began taking hits from the VC fire and had to pull pitch and take off for an emergency landing on the Island. With the remaining lift ship, it was decided to try and take a different approach to the LZ. I had the remaining lift ship fly very low-level downstream over the Bassac River above the LZ at approximately 5 feet above the water. They then did a pop-up and then down into the LZ. It avoided the VC fire and allowed the lift ship to pick up the Navy SEAL Team, including some seriously wounded. They then headed upstream on the Bassac River toward Binh Thuy Air Base and Hospital. En route, the lift ship AC informed me that the wounded SEALs were in bad shape, with one's head badly damaged. Later, in the SEAL Team debriefing, it was indicated that one of the Navy SEALs was new to the Team and had accidentally set off a booby trap which alerted the VC before the SEALs could attack their POW Camp.

After being treated at the Binh Thuy Air Base Hospital, the SEAL Team was taken to Okinawa for recovery. This ended the 13th CAB support for the Navy SEALs. However, after relocating from Soc Trang AAF to Can Tho AAF in November 1970 and transferring the Soc Trang AAF to the VNAF and their U.S. Air Force advisors, the Navy SEAL Platoon Leader visited my S-3 Office in Can Tho AAF. He and another Navy SEAL brought several cases of beer and boxes of steaks and said, "Thanks for saving my guys and getting them to the Hospital."

My Significant Activities & Stories in July—Nov 1970 at Soc Trang AAF

Fire Breakout in the Retrograde Ammunition Storage Area at SOS Trang AAF

As shown in the picture of Soc Trang AAF in Figure 34, the 13th CAB Retrograde Ammunition Dump was on the far side of the Airfield. A fire was initiated in the dump by a VC mortar fire that threatened to destroy aircraft and a possible follow-on attack by the VC. When the fire engine broke down en route, I aided in the resupply of fire extinguishers to the area, continuously driving through exploding ordnance to get the equipment to the men at the scene. I had to fight stray fires several times to stop them from spreading to critical areas. Without my coordination of resupply, the other firefighters would not have been able to obtain the necessary equipment. I was awarded the Soldiers Medal for these actions. It stated that through my leadership abilities and incredible display of courage, I was instrumental in extinguishing the spread of dangerous flames. It said that my actions were in keeping with the highest traditions of military service and reflected great credit upon myself, my unit, and the United States Army.

Award of The Distinguished Flying Cross Award for Heroism While Participating in Aerial Flight Evidenced by Voluntary Actions Above and Beyond the Call of Duty

Less than a week after the 13th CAB Retrograde Ammunition Fire, I took part in the 13th CAB TOC operations. A friendly unit was making insertions when a 164th CAG helicopter was shot down ten-miles miles northeast of Phung Hiep. It was in a heavily wooded area infested with enemy troops. I acknowledged notification of the downed aircraft and contacted the 13th CAB instructor pilot, CW3 Ruzzene, and we flew immediately from Soc Trang to the enemy-infested crash site. When we got close, I flew low-level and nap-of-the-earth (NOE) to identify the exact location of the downed aircraft. I carefully maneuvered the aircraft and landed in an insecure rice paddy next to the wooded area, barely large enough for one aircraft. We were informed by the friendly unit that a VC unit was proceeding by sampans to our location. While I kept the aircraft running, I instructed CW3 Ruzzene to move into the heavily wooded area, check out the downed aircraft for any survivors and bring them back to the aircraft as soon as possible.

When he arrived at the downed aircraft, CW3 Ruzzene informed me that the crew members were all dead and the area was relatively secure, although the VC were in the area. I told CW3 Ruzzene to bring what he could and return to the aircraft for a quick departure as another aircraft from the 164th CAG Headquarters with their flight surgeon was incoming and planned to land in our only feasible LZ for one aircraft. We then flew back to Soc Trang. I was informed that evening that the flight surgeon and pilots were being put in for Silver Stars for their heroic actions.

I then nominated CW3 Ruzzene for a Silver Star for his bravery and actions, which he later received. I was nominated and then awarded a Distinguished Flying Cross (DFC) for my conspicuous bravery and devotion to duty in keeping with the highest traditions of military service and reflected great credit upon myself, my unit, and the United States Army. Figure 47 is shown me receiving the DFC from the 164th CAG Commander at Can Tho AAF.

Figure 47. CPT Schrage Receiving DFC at 164th CAG HQs Can Tho AAF

In Country Evaluation of OH 58A and OH 6A Scout Helicopters in Combat Operations with the C-Troop 16th Cav Based on LTC Sauers 13th CAB Commander request

Background on the Army Light Observation Helicopters (LOH) Procurement in the 1960s

In the 1960s, the https://en.wikipedia.org/wiki/United_States_ Army the **United States Army issued** Technical Specification 153 for a LOH, capable of fulfilling various roles: personnel transport, escort and attack missions, casualty evacuation, and observation. Twelve companies took part in this competition.

Hughes Tool Company's Aircraft Division submitted the *Model 369.* Two designs submitted by Fairchild-Hiller and Bell were selected as finalists by the Army-Navy design competition board. However, the U.S. Army later included the Model 369 helicopter from Hughes.

During the competition, the Bell submission, the YOH-4, was eliminated as being underpowered (it used the 250 shaft horsepower(shp) (186 kW) T63-A-5). The bidding for the LOH contract came down to Fairchild-Hiller and Hughes. Hughes won the competition, and the Army awarded a contract for production in May 1965, with an initial order for 714, later increased to 1,300 with an option on another 114.

Hughes' price was $19,860 per airframe, less engine, while Hiller's price was $29,415 per airframe, less engine. The Hiller design (designated OH-5A) had a boosted control system, while the Hughes design did not. This would account for some of the price difference.

Howard Hughes, Hughes Helicopters CEO, is reported to have told Jack Real, his LOH VP, that he lost over $100 million in building 1,370 airframes. It was reported that Howard Hughes had directed his company to submit a bid at a price below the actual production cost of the helicopter to secure this order. This resulted in substantial losses to Hughes Helicopters in the U. S. Army deal, with the anticipation that an extended production cycle would eventually prove financially viable.

In 1968 Hughes submitted a bid to build a further 2,700 airframes. Stanley Hiller, Hiller Aircraft CEO, complained to the U.S. Army that Hughes had used unethical procedures; therefore, the Army opened the contract for rebidding by all parties. Hiller did not participate in the rebidding, but Bell did with their redesigned Model 206. After a competitive fly-off, the Army asked for sealed bids. Hughes bid $59,700 per airframe, while Bell bid $53,450. Reportedly, Howard Hughes had consulted at the last moment with his confidant Jack Real, who recommended a bid of $53,400. Hughes added $6,000 to the bid without telling Jack Real and thus lost the contract to Bell, whose bid was $53,450. The OH-6A Scout was an excellent LOH aircraft in South Vietnam. (*The Asylum of Howard Hughes*, Jack Real with Bill Yenne)

In 1964 the U.S. Department of Defense issued a memorandum directing that all U.S. Army fixed-wing aircraft be transferred to the U.S. Air Force while the U.S. Army made the transition to rotor-wing aircraft.

The U.S. Army's fixed-wing airplane, the **O-1 Bird Dog**, which was utilized for artillery observation and reconnaissance, would be replaced by the OH-6A helicopter. The aircraft entered service in 1966, arriving in the **Vietnam War**. The pilots dubbed the new helicopter *Loach*, a word created by the pronunciation of the acronym of the program that spawned the aircraft, LOH.

Shortly after production began, the OH-6 started to demonstrate its impact on the world of helicopters. The OH-6 set 23 **world records** for helicopters in 1966 for speed, endurance, and time to climb. On 26 March 1966, Jack Schwiebold set the closed-circuit distance record in a YOH-6A at **Edwards Air Force Base**, **California**. He flew without landing for 1,739.96 mi (2,800.20 km). Subsequently, on 6 April 1966, **Robert Ferry**, Hughes test pilot, set the long-distance world record for helicopters. He flew from Culver City, **California**, with over a ton of fuel to Ormond Beach, **Florida**, covering 1,923.08 **nm** (2,213.04 mi, 3,561.55 km) in 15 hours, and near the finish at up to 24,000 feet (7,300 m) altitude. As of 2018, these records still stood.

In 1967, the Army reopened the LOH competition for bids, and Bell resubmitted for the program using their Model 206A design. Fairchild-Hiller failed to resubmit their bid with the YOH-5A, which they had successfully marketed as the FH-1100. In the end, Bell won the contract, and Model 206A was designated as the OH-58A. Following the Army's naming convention for its aircraft, the OH-58A was named **Kiowa** for the Native American tribe.

Figure 48A. Hughes OH-6A Cayuse *Figure 48B. Bell OH-58A Kiowa*

Shown in Figure 48A is the OH-6A Cayuse, winner of the first LOH competition, while the winner of the second LOH competition was the OH-58A Kiowa, as shown in Figure 48B

Almost all the Air Cav troops loved their OH-6A Loaches. As a result, all except one Air Cav Troop in Vietnam refused to turn in their OH-6As for OH-58As. LTC Robert Sauers, the 13th CAB Commander, had been involved with the second LOH competition and was very upset with the selection of the OH-58A over the OH-6A. As a result, he tasked me with doing an in-country operational evaluation of the S-3 OH-58A with a C Troop, 16th Cav OH-6A Scout aircraft, as part of the Scout-Attack Team illustrated in Figure 49. Since I was not checked out to fly an OH-6A, I was sent to Vung Tau for the one-week OH-6A transition course.

Scout-Attack Team

OH–58A   OH–6A Cayuse   AH–1G Cobra

Figure 49. Aircraft for Evaluation in the Scout-Attack Platoon

As illustrated in Figure 3, Vung Tau served as an in-country rest and relaxation (R&R) center for U.S. and South Vietnamese. Some say the VC used it for similar purposes. A picture I took from my OH-6A training helicopter of the City of Vung Tau is in Figure 50A, and the OH-6A helicopter I transitioned in is shown in Figure 50B. I enjoyed my week in Vung Tau learning to fly the OH-6A. I also spent some time with my former USMA West Point roommate, Dean Hanson, assigned as the Army's Vung Tau Mayor.

Figure 50A. Picture of Vung Tau City

Figure 50B. OH-6A Training Helicopter

THE EVALUATION OF SCOUT HELICOPTERS WITH C-TROOP, 16TH AIR CAV

As mentioned, C Troop was reactivated in Vietnam using assets of D Troop, 1/4th Cavalry, which had been in the country with the 9th Infantry Division. It served with the 13th CAB from March 1970 until February 1973. I flew and evaluated the OH-6A and the OH-58A for agility & maneuverability at the request of the 13th CAB Commander, LTC Bob Sauers. I provided a report which touted the advantages of the OH-6A Cayuse over the OH-58A Kiowa.

Interestingly enough, in 1978, when I was the Aeromechanics Branch Chief in AVRADCOM, I was selected to be the Technical Director for the Army Helicopter Improvement Program (AHIP) SSEB for a new scout helicopter from a derivative aircraft. It turned out to be the third competition between an upgraded OH-6 versus an upgraded OH-58. In this competition, the derivative OH-58 with a new main rotor system turned out to be a better selection. It was eventually fielded as the OH-58D Kiowa Warrior

TRAINING VNAF PILOTS AND TRANSFER OF SOC TRANG AFF TO VNAF AND USAF

In September 1970, we trained VNAF pilots on airmobile operations out of Soc Trang AAF. Most pilots came directly to Soc Trang from U.S. pilot training at Fort Wolters, TX, and Fort Rucker, AL, USA. As the Asst S-3/S-3 for the 13th CAB, I was responsible for setting up the operational training for the VNAF "Peter Pilots" (new pilots) with ACs in the 121st AHC and 336th AHC. It was part of the Helicopter Vietnamization Program at Soc Trang AAF, and lift ships from the 121st AHC and 336th AHC were transferred to the VNAF and USAF in early November 1970.

The 121st AHC and 336th AHC ACs were not very warm about flying with new VNAF pilots in combat. I remember giving them a pep talk when the first batch of VNAF pilots arrived on a Friday before starting the operational combat training on Monday. I introduced the approximately 20 VNAF pilots to the ACs they would be flying with on the following Monday. Only one VNAF pilot showed up. After investigation, I found out that the other VNAF pilots went back to their homes to visit their families and girlfriends over the weekend. They had been gone for the nine-month U.S. pilot training, and they were homesick. My credibility with the ACs was greatly diminished. Fortunately, we could get all of the VNAF pilots back by Wednesday to begin the operational training.

Figure 51. LTC Sauers, CO 13th CAB tranfers 121st and 336th Helicopter to VNAF and his Change of Command Ceremony

The lift ships from the 121st AHC and 336th AHC were transferred by LTC Robert Sauers, CO 13th CAB, to the VNAF and USAF in October 1970 as illustrated in Figure 51. LTC Sauers change of command ceremony took place in Can Tho in December 1970. He was a great commander and served as my mentor. I will never forget his confidence in making me his S-3.

The official transfer and ceremony for transferring the Soc Trang AAF and lift helicopters from the 121st AHC and 336th AHC to the VNAF was a significant milestone for the Helicopter Vietnamization Program. Once again, being the 13th CAB S-3/Asst. S-3, I had substantial responsibility in organizing the transfer. It even included getting the 21st ARVN Division Band to provide the music for the transfer. U.S. and South Vietnam national and military leaders were present at the Official transfer on November 4, 1970.

Figure 52A. U.S. Secretary of State William Rogers.

I am shaking hands with U.S. Secretary of State William Rogers in Figure 52A. I had briefed Secretary Rogers on the status of Helicopter Vietnamization the day before. Secretary of Air Force Seamans and General Abrams are approaching in Figure 52B. GEN Creighton Abrams, U.S. military commander in Vietnam, was there as Vietnamization had been one of his key initiatives.

Figure 52B. Approaching Dignitary

Figure 53A. VIPs on Reviewing Stand

*Figure 53B. VIPs Reviewing Transfer
w/21st ARVN Band Leader.*

VIPs on a stand for the Transfer Ceremony, included from right to left in Figure 53A, are Dr. Robert C. Seamans, Secretary of the Air Force; GEN Creighton Abrams, U.S. Army Vietnam Commander; GEN. Lucius Clay, 7th Air Force Commander, and Vietnamese GEN. Minh, CG 21st ARVN Division. Shown in Figure 53B is the Reviewing Party for the Transfer and the Band Leader for the 21st ARVN Division Band, who I had convinced to bring his band to this Transfer Ceremony.

Figure 54A. Saluting the Transfer of Soc Trang AAF.   *Figure 54B. Transferring the UH-1s to VNAF*

The official Transfer turned over Soc Trang AAF to the RVNAF or VNAF. Concurrently, the U.S. Army 121st AHC and 336th AHC transferred their aviation and support assets to the VNAF. 31 UH-1 helicopters were involved in the turnover, the fourth Transfer of a complete helicopter company to the VNAF in the previous two months. Receiving the aircraft were the newly activated VNAF 225th and 227th Helicopter Squadrons. In addition, some personnel of the U.S. units remained at Soc Trang, assisting the VNAF during the transition period. The transfers of the airfield and aircraft were part of the titled RVNAF Improvement and Modernization Program, generally termed as Helicopter Vietnamization. Pictured in Figure 54A are MAJ Melton and CPT Schrage saluting the AAF Transfer, and in Figure 54B, CPT Schrage passes by the VNAF pilots in front of the transferred UH-1 aircraft.

* * *

ACTIVITIES AND MY STORIES OUT OF OF
CANTHO AFF NOV 1970—JANUARY 1971

MY STORIES FROM NOVEMBER 1970 TO
JANUARY 1971 AND RETURN TO THE USA

The Transfer of Soc Trang AAF and its aviation assets to the VNAF and Air Force advisors on November 4, 1970, resulted in the 13th CAB moving its headquarters to Can Tho AAF. There was concern about whether the VNAF could provide the necessary support to the 21st ARVN Division. As mentioned earlier, there was no love lost between the VNAF and ARVN, especially with the 21st ARVN Division.

It also appeared to the 13th CAB S-3 Staff that the USAF Advisors were more interested in human comforts, such as air conditioning and television reception, than they were in building rapport between the VNAF and the 21st ARVN Division. Also, there seemed to be more VNAF pilots than the necessary mechanics required to keep up the necessary operational availability for their helicopters.

When I returned from South Vietnam and attended the U.S. Army Field Artillery Officers Advanced Course, I wrote an essay on *"Helicopter Vietnamization – Would Work or Not."* My conclusion was that it would not. It turned out to be the winning essay out of 150 papers submitted by the other Advanced Course students.

During the period 1970 to 1971, the U.S. was beginning to draw down its combat forces, and the new watchwords were Vietnamization & Withdraw. It was the period when the will of the U.S. to prosecute the war had slipped, and the U.S. was transferring responsibility to the South Vietnamese as the only remaining hope for victory. This was a critically trying period for U.S. Army advisors to South Vietnamese ARVN, Ranger, and Airborne units. As the U.S. combat units withdrew, the units they advised spearheaded several campaigns in South Vietnam, Cambodia, and Laos. Often outnumbered and outgunned, the elite Ranger and Airborne units fought Viet Cong and North Vietnamese units in some of the most challenging terrain in Southeast Asia, ranging from the legendary U Minh Forest and Seven Sister Mountains in the Mekong Delta to the rugged hills of Southern Laos.

In January 1971, the 7/1 Armored Cav Squadron (AC.), in a particular operation, supported the 9th ARVN Division in reaching a beleaguered U.S. Special Forces Team on Ta Bec Mountain, one of the Seven Sister Mountains. U.S. troops of "D" Troop, 7/1 ACs were inserted to assist in the rescue. This is noted as the last time U.S. Troops were employed in the Mekong Delta.

In April 1971, elements of the 164th CAG engaged in decisive action against NVA units infiltrating from Cambodia to the U Minh Forest. ARVN ground troops were used exclusively and gave a good account of themselves in some of the most hostile actions to date.

TRANSFER OF 13TH CAB HQS AND S-3 SHOP TO CAN THO AFF

The relocation of the 13th CAB HQs and S-3 Shop went smoothly. The offices and quarters were good and not the tent city that the 162nd AHC still occupied. I continued to fly my OH-58A Kiowa as a command-and-control ship or for special ash and trash missions. As mentioned earlier, the Navy SEAL Platoon Leader whom we had supported in attempted POW rescue operations visited my S-3 Office, bringing cases of beer and steaks, to thank me for getting his SEAL Team somewhat intact out of a bad situation.

AN ASH AND TREASH MISSION TO CORPUS-CHRISTI BAY SHIP THAT ALMOST WENT AWRY

The U.S. had over 12,000 helicopters in Vietnam. With so many helicopters, transporting the damaged ones back to the United States for repairs would've been a logistical nightmare. So, instead of bringing helicopters to the repair facility, America brought the repair facility to the helicopters in the form of the USNS Corpus Christi Bay (Ship) (Figure 55A). From 1966 to the end of the Vietnam War, USNS Corpus Christi Bay served as a floating repair depot for helicopters.

USNS Corpus AAF urgently needed a helicopter part that could only be obtained from the USNS Corpus Christi Bay repair facility in the South China Sea. So, I volunteered to fly a mission in my OH-58A as a single pilot to the USNS Corpus Christi Bay to get the part. Damaged choppers were brought in by barge, fixed, and returned to the front lines.

During the monsoon season and other times during the year, a thick cloud/fog cover would cover a good portion of the Mekong Delta. Since most helicopter pilots, such as myself, did not have a full IFR Certificate, we would often fly up through and above the clouds and then look for a hole in the clouds to descend to the area close to our destination. The only navigation aid we usually had was a Nondimensional Directional Beacon (NDB). An NDB is a radio beacon operating in the M.F. or L.F. bandwidths. NDBs transmit a signal of equal strength in all directions, containing a coded element used for station identification (usually 1-3 letters in Morse Code). NDBs are often associated with Non-Precision Approach procedures.

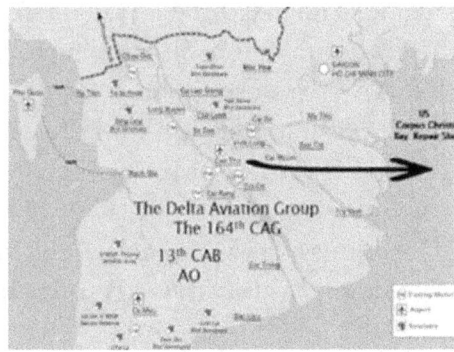

Figure 55A The Corpus Christi Bay Ship   *Figure 55B My Route from Con Tho to Ship*

Unfortunately, the cloud cover over the Mekong Delta that day, especially in the direction I was heading toward, did not have any holes. I finally found a hole that was over the South China Sea. The NDB was wavering back and forth, about + or - 20 degrees, toward the direction of USNS Corpus Christi Bay. As I dropped down through the hole into VFR, I noticed I had a 10-minute fuel warning. I continued to try and follow the NDB direction toward the ship. After flying for about five more minutes without seeing the ship, I considered ditching the aircraft. However, I had little training or experience in ditching an aircraft, especially in the ocean. Just when panic started to set in, I saw the ship and proceeded to land, get the critical part, refuel, and fly back to Can Tho AAF. Much of the cloud cover had dissipated, and I had a complete VFR flight back to Can Tho. It was the second time I almost bought the farm while flying in South Vietnam.

Iroquois Night Fighter and Night Tracker (Infant) Technology in the U-Minh Forest.

In early 1970 the US Army Concept Team in Vietnam evaluated the INFANT to determine its combat suitability for stability operations in RVN. It was initially installed on UH-1M helicopters; a utility helicopter converted from US Army UH-1Cs to an attack helicopter during the Vietnam era. The UH-1M was upgraded from the UH-1C models with a more powerful Lycoming T53-L-13B 1400 shaft horsepower (shp) engine. The first three UH-1Ms were equipped with the INFANT system for night operations. Beginning in late 1970 and early 1971, the UH-1M INFANT was used by the 13th CAB to intercede NVA and Viet Cong insertions into the U Minh Forest and along the Cambodian Borders out of Moc Hoa AAF. Much of the NVA insertion into the U Minh Forest was along the coast of the Gulf of Siam, using fishing ships as surrogates. These missions were successful for the ARVN 21st Division Combat Operations with 13th CAB support in Spring 1971.

Rest and Relaxation in Honolulu, Hawaii, the Last Week in December 1970

After an almost year of combat activities in South Vietnam, a one-week rest and relaxation (R&R) in Hawaii was more than welcome. Fortunately, my wife Nancy was able to join me after giving birth to our daughter, Susan, in November 1970. Our oldest son, Steven, was born in March 1969 before I returned from serving in Germany. Both Steven and Susan were born at Scott AFB Hospital in Shiloh Valley, Illinois. In addition, I am thankful that I was in the States for the birth of my last two

Figure 56A. Nancy & Dan at Don Ho Restaurant.   *Figure 56B. Dan Looking Out Illikai Hotel*

children, Michael and Alex. A picture of Nancy and me in Honolulu is shown in Figure 56A. A view of me looking out at the Rainbow Towers Hotel is shown in Figure 56B.

We definitely enjoyed our week, taking in a Don Ho Concert at his restaurant, the Ike and Tina Show at the Rainbow Towers Hotel, and a road trip around Oahu Island.

RETURN TO THE SOUTH VIETNAM AND MY LAST MONTH IN THE COUNTRY

I returned to South Vietnam from R&R on January 2, 1971, with mixed emotions for my last few weeks in the Country. While we had accomplished a lot in the Mekong Delta with the 13th CAB and the 164th CAG, along with the ARVN divisions, I knew that Cambodia and South Vietnam were in big trouble. Helicopter Vietnamization and Vietnamization, in general, were not going to work. I kept seeing those teenagers in the Cambodian Khmer Army when we landed in the Elephant Mountains who thought we were coming to save them and their country.

I believe we missed a critical opportunity to make it a "Better War" if a coalition of GEN Lon Nol's Cambodia with South Vietnam with US Air Support could have been formed and implemented. I believe a book on setting the record straight, as called for by GEN Schwarzkopf, is still required and supports the *clear, hold* and *build* idiom from the book *A Better War*. The book should assess the Vietnam War and the attempted Iraqi and Afghanistan nation-building efforts as well.

I transitioned my 13th CAB S-3 Shop to the new S-3 and tried not to participate in combat operations. There was a tradition for short-timers in the 13th CAB to go to the MACV Hotel in downtown Can Tho and get a suntan on the rooftop before departing for the States. As it turned out, the one day I tried it, the hotel was being mortared by the Viet Cong. So, I decided it was wise to stay on base in the 13th CAB S-3 Shop until it was time to fly back to the States.

"THE WHY" THAT THE BOOK HOPES TO ACCOMPLISH

I believe I have established in this chapter of Book 1 in my Trilogy, the foundation for the "Why" of this book. I will build on it in the follow-up chapters and books to set the record straight and tell the rest of the Trilogy story: *A Full Lifetime Career of Seeking Perfection Driven by Family and Mentors.*

Chapter Three

Growing Up and Moving Forward

"Country roads, take me home To the place I belong"- John Denver

Growing Up in the Midwest

I was born shortly after midnight on March 18, 1944, in the Red Bud, Illinois Hospital. It was the closest hospital to our home in Prairie du Rocher, Illinois, as shown in Figure 57. Other small towns on country roads where I lived while growing up are also illustrated with an *. Also identified with double asterisks** are key locations and rivers which will be discussed, such as the high school, Mater Dei High School in Breese, Illinois, from which my wife, Nancy, and I attended and graduated. The town we lived in for ten years, O'Fallon, Illinois, our home from 1974-84, is identified and will be discussed in Book 2. Also, the Illinois, Kaskaskia, Mississippi Rivers, and Carlyle, where Nancy grew up, and its Lake are identified. We grew up in the small towns linked by the Kaskaskia and Mississippi River Basins, the key agriculture and water supply sources in Southern Illinois. They are described in some detail.

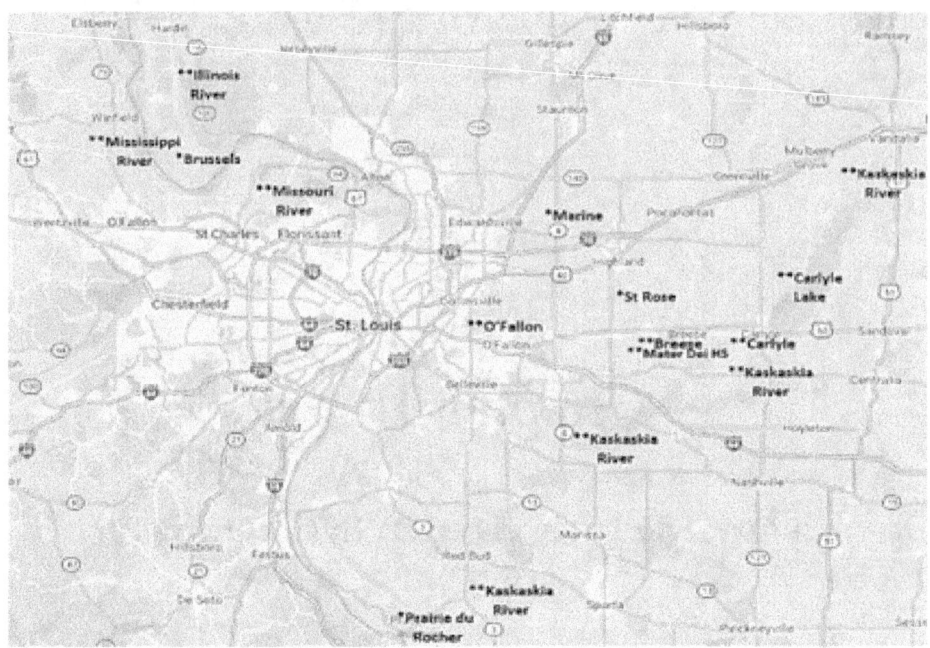

Figure 57. The small towns* in which I lived growing up in Southern Illinois around St. Louis, MO

Figure 58A. The Albin J. Schrage Family.

A picture of my family; father Albin, mother Mary, twin sisters Jeanne and Joan, my middle sister Juanita, and I are shown in Figure 58A. A later picture of the children is shown in Figure 58B.

My father, Albin Joseph, and mother, Mary Helen, were schoolteachers. My father taught mostly high school, and my mother taught grade school. As illustrated in Figure 57, we moved quite a bit as my father and mother sought better teaching and higher-paying jobs in different towns. In addition to teaching high school courses, my father also coached high school and grade school baseball and basketball. In addition to her teaching, my mother was a perfectionist about housekeeping – our house was always immaculate.

My father was of German descent and grew up on a dairy farm near St. Rose, IL (See Figure 57). He had three older brothers and four sisters, most of whom became or married dairy farmers. My father was also the only one who attended college, with two years at Illinois State Teaching College in Normal, Illinois. He taught at Linden Grove

Figure 58B. The Albin J. Schrage Family Children

Grade School, near Breese, IL, for ten years and each summer attended college to complete his bachelor's degree in education. My mother was mostly of Irish descent, with a little bit of Cherokee Indian descent. Her grandmother, an escapee from the "Trail of Tears," married her grandfather and settled in Southern Illinois. She had a sister and two brothers. Her mother, Mary B. McQuade, was a teacher and became the first woman Clinton County Illinois Superintendent of Schools. My mother initially attended a Catholic convent in Ruma, Illinois, not far from Prairie du Rocher, following her sister, Irma, but did not finish. She earned her teaching certificates from several colleges before graduating from McKendree College in Lebanon, IL. Both of my parents were devout Catholics. If I had been born a few hours earlier on St. Patrick's Day, my first name would have been Patrick instead of Daniel. My first home, Prairie du Rocher, Illinois, was a small town with a fascinating history. As can be seen from Figure 51, it was south of St. Louis, MO, and very close to the Mississippi River.

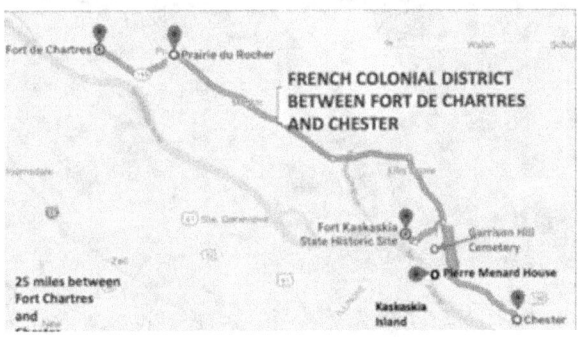

Figure 59A. French Colonial Period.

Figure 59B. Prairie du Rocher
and American Bottom

Fig. 60A. Aerial View Prairie du Rocher

Fig. 60B. Picture from Bottom to Town and Church

Figure 60C. Picture of Bluff Road.

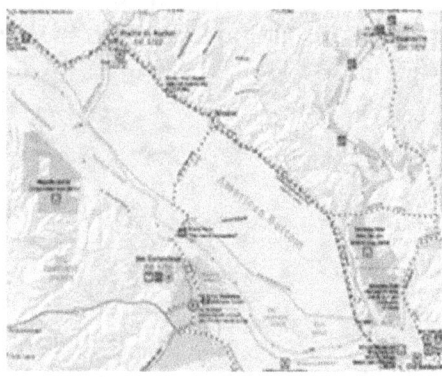

Figure 60D. Bluff Road along American

PRAIRIE DU ROCHER, ILLINOIS

In 1943, my father left Linden Grove School to become a Prairie du Rocher Grade School principal. Prairie du Rocher is a village in Randolph County, Illinois, founded in the French colonial period in the American Midwest (Figure 59A). The French Colonial District included Fort de Chartres, Prairie du Rocher, Fort Kaskaskia, and Chester. The Prairie du Rocher community is located near bluffs that flank the east side of the Mississippi River along the floodplain. They are called the "American Bottom," Figure 53B.

An Aerial View of the American Bottom with crops and Prairie du Rocher is shown in Figure 60A. A photograph from the American Bottom to Prairie du Rocher and St. Joseph's Catholic Church, one of the oldest churches in the USA, is shown in Figure 60B. Founded in 1721, it celebrated its 300th anniversary in 2021 and the town of Prairie du Rocher in 2022. It has averaged about 600 residents. Pictured in Figure 60C is the Bluff Road that runs parallel to the bluffs from Prairie du Rocher and the Mississippi River, as illustrated in Figure 60D.

Floods have been a significant problem for Prairie du Rocher over the years. When my father was teaching there, I remember stories about the whole town packing sandbags, including my father, to keep the town from being flooded. In 1993 Prairie du Rocher was one of the few towns along the Mississippi River that escaped flooding in the Great Flood of 1993. After levees broke to the north near Columbia and Valmeyer, Illinois, floodwaters engulfed Fort de Chartres and threatened the town of Prairie du Rocher. The village called for help, and its citizens added a foot and a half of sandbags to the creek levee to keep the town from flooding. As illustrated in Figure 61, townspeople today continue to practice packing sandbags to prevent future floods. (*The* Prairie du Rocher | Modoc | Edgar Lakes Levee District Strategic Plan).

Figure 61. Prairie du Rocher Town People Packing and Installing Sandbags

FORT KASKASKIA, KASKASKIA VILLAGE, AND KASKASKIA ISLAND

The key fort/village/island that emerged in the French Colonial Period is identified in Figure 59A. Kaskaskia Village was an important

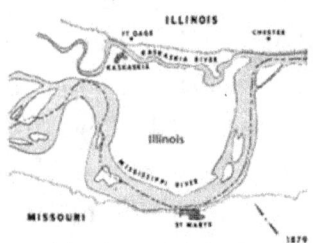

Figure 61A. Mississippi
Chanel before 1881.

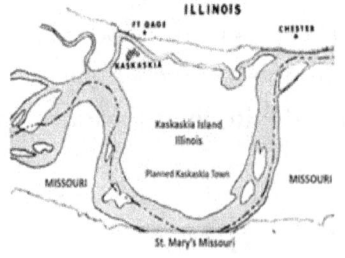

Figure 61B. The merging
of the Kaskaskia River with
the Mississippi River

French colonial town with a peak population of over 7,000 residents. Kaskaskia then became the capital of the Illinois Territory and later the first state capital of Illinois in 1818 winning a close election with Carlyle, Illinois. The town was home to leading political and economic figures in the early shaping years of Illinois. Good fortune, however, did not long endure. Natural disasters of unprecedented magnitude plagued the town and robbed its vitality. People came to regard Kaskaskia, once the center and focus of Illinois, as just a quaint and somehow foreign relic. In 1881, the town was destroyed when the Mississippi River shifted to a new channel combining with the Kaskaskia River, resulting in Kaskaskia Island.

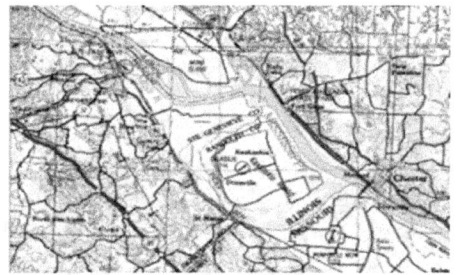

Fig. 61C. Current Map of Integrated

Figure 61D. Rebuilt Town on Kaskaskia Island

The townspeople rebuilt but again had to be evacuated during the Great Flood of 1993. Now, about nine people live there. The shifting of the Mississippi River, before and after forming a new combined Mississippi and Kaskaskia River Channel was formed, as illustrated in Figures 61A and 61B. Figure 61A shows that the Kaskaskia River meandered close to the Mississippi River but did not connect. Through erosion and tree removal, the Kaskaskia River joined with the Mississippi River to form a new channel (Figure 61B). A remaining channel of the Mississippi River followed the previous channel, which created Kaskaskia Island. Over the years, the Kaskaskia River flowed directly into the Mississippi River, forming the main channel with a secondary channel around Kaskaskia Island. This created a more current flow in Figure 61C which illustrates an updated map of Kaskaskia Village on Kaskaskia Island. It shows that the Kaskaskia River has merged with the Mississippi River as the main channel, making it easier to transport cargo from one river to the other. In 1893 the people of the town moved and rebuilt the Church of the Immaculate Conception at Kaskaskia. They also built a shrine in a similar style nearby to house the "liberty bell." Illustrated in Figure 61D is a picture of the rebuilt Kaskaskia Village on Kaskaskia Island.

Kaskaskia Island is the only piece of Illinois west of the Mississippi with St. Mary's Missouri across the smaller channel of the Mississippi River. In 1993 the Mississippi River almost completely flooded the island, as illustrated in Figure 62A.

As of the census of 2000, there were nine people, four households, and three families residing in the village. The population density was 83.0/sq mi (32.0/km²). There were five housing units at an average density of 46.1/sq mi (17.8/km²). The racial makeup of the village was 7 White, 1 Pacific Islander, and one from other races. There were 2 Hispanics or Latinos of any race. Several years ago, Figure 62B, a Welcome Kaskaskia sign, listed the population as 18.

Figure 62A. 1993 Kaskaskia Village Flooded.

Figure 62B. Village of Kaskaskia Sign 1

KASKASKIA RIVER BASIN AND ITS REACH INTO SOUTHERN ILLINOIS TOWNS WHERE WE LIVED

The Kaskaskia River Basin and the Illinois River Basin are illustrated in Figure 63A. The Kaskaskia River Basin is the second largest river basin in Illinois, next to the Illinois River Basin. In Figure 63B, the Kaskaskia River Basin is broken down into the Silver Creek Watershed, the Plum Creek Watershed, and the Carlyle Lake.

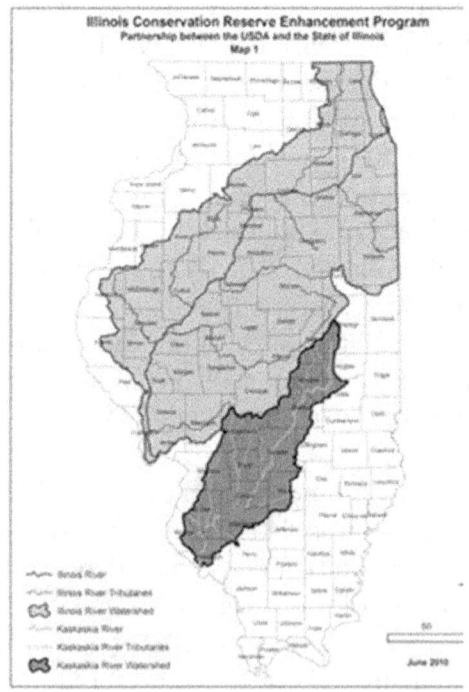

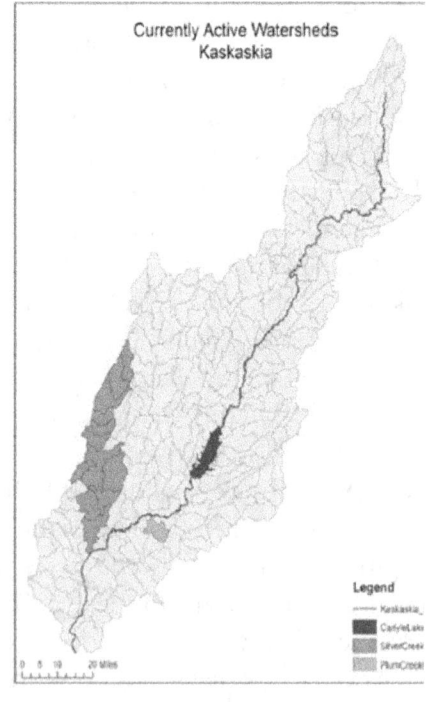

Figure 63A. The Illinois and Kaskaskia River Basins.

Figure 63B. Breakdown of Kaskaskia River Basin

Figure 64A. Picture Representative of Silver Creek Watershed.

Figure 64B. Plum Creek Watershed

A picture of the Silver Creek Watershed is shown in Figure 64A. A picture of a Coal Loading Station next to the Kaskaskia River and the Plum Creek Watershed is shown in Figure 64B. Finally, views of Carlyle Lake are shown in Figure 65. These watersheds are critical for farming and agriculture in Southern Illinois.

Referring back to Figure 57, the small towns in which we grew up in the Kaskaskia River Basin and its watersheds, touched many of these small towns and the other identified locations. These will be discussed later in this chapter and in subsequent chapters.

MARINE ILLINOIS

My family's next stop was Marine, Illinois (See Figure 57). After two years at Prairie du Rocher, my father was appointed principal of the Marine Grade School and two-year High School, where we remained for four years. He also coached the basketball and baseball teams. Marine was another small town with country roads sand fewer than 1000 people. Its location and an aerial view are shown in Figures 66A and 66B.

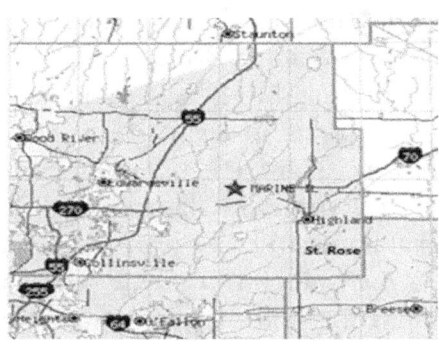

Figure 66A. Map of Marine, IL Area

Figure 66B. Aerial View of Marine, IL.

BRUSSELS, ILLINOIS

EXPERIENCES AS A PSEUDO "TOM SAWYER" OR "HUCK FINN"

After Marine, IL, my father accepted an assistant principal, teaching, and coaching position at a four-year high school, Brussels High School in Calhoun County, as illustrated in Figures 57 and 67.

As can be seen, Brussels, Illinois, is found at the lower end of a peninsula from the confluence of two rivers, the Mississippi and Illinois. The peninsula is about seven miles wide and must be accessed by ferry – the Brussels or Grafton Ferry if coming from Illinois, or the Golden Eagle Ferry or the Batchtown-Winfield Ferry if crossing the Mississippi River from Missouri. The only other access to Brussels, other than ferry boats, is the Hardin Bridge about 30 miles north, as illustrated in Figure 64. Ice and flooding can shut the ferries down during the winter and spring, making Brussels isolated from the rest of the world transportation-wise. This provided a unique experience, akin to Tom Sawyer and Huck Finn, growing up on the Mississippi River in Mark Twain's books.

An aerial picture of Brussels, Illinois, is shown in Figure 68A. It was the smallest town in which the Schrage family lived. A welcome sign shows a population of 150, as shown in Figure 68B. I don't believe the population has fluctuated much over its lifetime.

The Schrage house shown in Figure 68A was the second of four houses that the Schrage family lived in during their seven years in Brussels, Illinois. The first house we lived in was behind the Wittman Hotel and needed substantial repair. The Wittman Hotel is pictured in Figure 69A. The Wittman Hotel, one of the most prominent buildings in town, still exists today as the Wittman Restaurant.

Shown in Figure 69B is St. Mary's Catholic Church, which the Schrage family were members for their seven years in Brussels. My best friends growing up in Brussels were the Imming brothers, Leland, and Larry, who lived down the street from our house in Figure 68A. They also had a television, which we did not have, so we spent considerable time after school watching television at their house. My father bought a television so we would spend more time at home. We also learned to fish and hunt using our bicycles to crisscross the Calhoun County Peninsula to access the best fishing holes and hunting locations. In many ways, we grew up as surrogates of Tom Sawyer and Huck Finn, who lived up the

Figure 67. Access to Brussels, Illinois by Ferry Boats or a Bridge

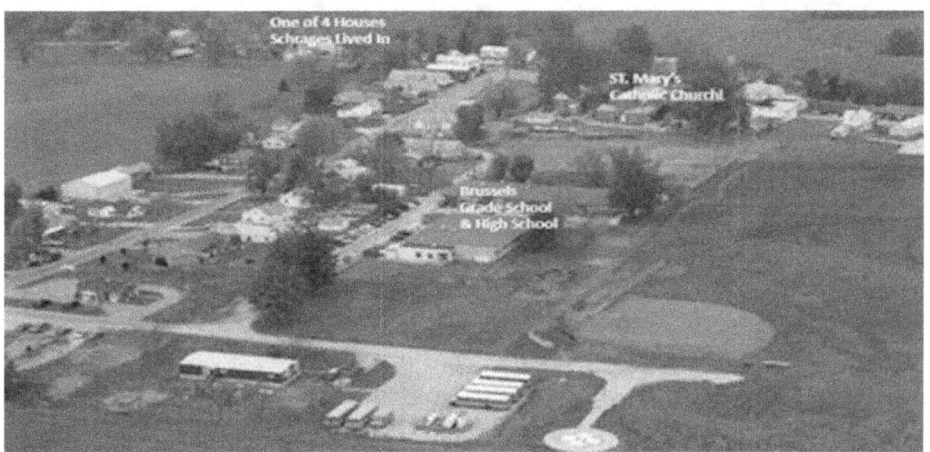

Figure 68A. Key buildings in Brussels, Illinois.

Figure 68B. The welcome and population sign.

Mississippi River in Hannibal, Missouri, in Mark Twain's novel.

The last two houses in Calhoun County in which we lived were both in the country. We had to move out of the second house in town, as its owner, a schoolteacher named Tony Siemer, was moving back to Brussels to teach. He eventually became the high school principal.

97

Figure 69A. Wittman Hotel.

Figure 69B. St. Mary's Church

Figure 70A. Bantam or Banti Roosters

The first country house was on the Menke Farm several miles out of town. It had a large apple and peach orchard on it. My middle sister, Juanita, and I would pick up the fallen, mostly rotten apples on the ground and get 25 cents per bushel for making cider, while professional pickers picked the apples and peaches off the trees.

The second country house we lived in was the Haug House. It was a large white house with a beautiful front porch just a few miles from town. I would shoot at birds in the tree with my BB gun off the front porch. Mister Haug fixed up the house for us, and it was a substantial improvement over our previous homes. My sister Juanita and I raised Bantam (or Banti) chickens and rosters. We also had a black Labrador dog named Blackie, and several cats.

Figure 70B. How to Train your Dog

The Banti rosters, illustrated in Figure 70A, were especially mean and would chase after you to peck at you with their beaks. A mean streak could also be said of Blackie, who was good with us but didn't like strangers and was never properly trained, as suggested in Figure 70B. When we made our next move to St. Rose, Illinois, we gave Blackie to a cousin who lived on a farm, but he still had his mean ways with strangers, and they had to get rid of him.

My three sisters and I learned to play musical instruments and were part of a Brussels Grade School or High School band. Juanita played the clarinet while I played the cornet. My older sisters, Jeanne and Joan, played in the high school band. Jeanne played the clarinet, and Joan played the drums.

My father coached the Brussels High School Basketball and Baseball teams, as shown in Figures 71A and 71B. As can be seen, Jeanne and Joan, along with their best friend, Roseann, were the cheerleaders. I got early exposure to both sports, and by the time we moved to St. Rose, Illinois, following the sixth grade, I began to excel somewhat in both sports.

Figure 71A. Brussels HS Basketball Team Figure

71B. Brussels HS Baseball Team

Following my sixth grade and my sister Juanita's eighth grade in Brussels Grade School, my father was offered the principal and a teaching position at the St. Rose Grade School in his hometown of St.

Rose, IL. My mother was also offered a teaching position there. It ended our unique but isolated stay in Brussels, Illinois, and Calhoun County. It ended somewhat my experience in Huck Finn and Tom Sawyer country, although I did continue to hunt and fish in the St. Rose countryside. My older twin sisters graduated from Brussels high school and worked in St. Louis, MO, first with Graham Paper Company and then with travel companies. They went on to tour around the world, with both eventually having their own travel agencies.

EPERIENCES AS A RISING ATHLETE IN BASEBALL AND BASKETBALL IN ST. ROSE, ILLINOIS

St. Rose is another small town on country roads in Southern Illinois, surrounded mainly by dairy farms founded and run by German emigrants. As shown in Figure 72, it is located between Breese, Illinois, five miles south, and Highland, Illinois, about nine miles north. It is in Clinton County, Illinois, identified in Figure 72. Route 50 is the main highway that runs from the U.S. East Coast to the West Coast. In my final year at USMA we were allowed to take an elective Geography course. My class project was tracing and discussing Route 50 from

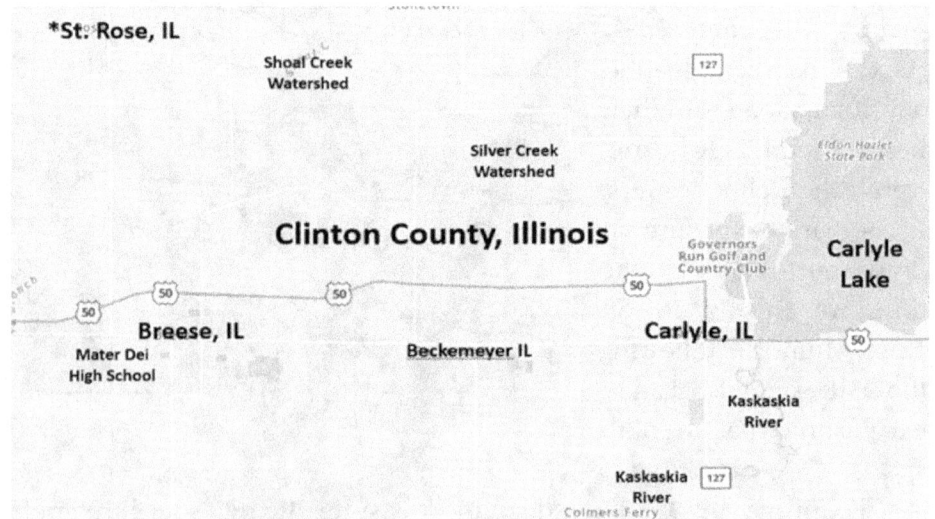

Figure 72. Key Towns and Locations in Clinton County, Illinois

Baltimore in the East to San Francisco in the West. Other cadets in the Class were from the East and West Coast so I got their insights. Overall, it was the most fun course I ever took. The key towns of Breese and Carlyle, Illinois, leading to Carlyle Lake, are shown. My wife, Nancy, was born and grew up in Carlyle, Illinois. We both attended and graduated from Mater Dei High School in Breese, Illinois. My sister, Juanita, did too and she worked with the Corps of Engineers developing Carlyle Lake before working for the government at Scott AFB, IL.

Figure 73. A picture of Main Street St. rose with key location identified

A picture of the main street through St. Rose and critical locations are shown in Figure 73. The Albin J. Schrage home is shown in our house location, although the original house had red brick siding. When we lived in it. Across the street was the Francis Woltering home where my uncle and aunt, Francis and Irene Woltering, my father's sister, lived with their five children: Eileen, Ruth Ann, Mary Edith,

Connie and David, who were our first cousins. David and I were best friends I remember we both played the cornet/trumpet and often orchestrated tunes or sounds across the street to each other.

Other locations named are the St. Rose Church grounds, which I mowed during most of my grade school and high school years, and the Schuette Bros Piggly Wiggly Store, where I worked as a shelf stacker and checkout clerk during my high school years. Another location is

*Figure 74A. St. Rose
Catholic Church*

Figure 74 B. St Rose School

Jung's Tavern, which was widow-owned and run with assistance from her daughters and son, Junior, Connie and David, remember we both played the cornet/trumpet and often orchestrated tunes or sounds across the street to each other who was also one of my best friends.

The other critical locations shown in more detail are the St. Rose Catholic Church, School, and Athletic Facilities, illustrated in Figures 74A and 74B.

My father was the St. Rose Grade School superintendent and taught the seventh and eighth grades. He also coached the basketball and baseball teams. He was an excellent teacher and coach and an avid sports fan who could easily quote area, college, and pro names and scores. A picture of the St. Rose Eighth Grade Class is shown in Figure 75, with me standing next to my father.

Figure 75. St. Rose Grade School 1958 8th Grade Class

Our basketball team was small and inexperienced. Plus, we had to play the larger private and public schools in Clinton County. I remember one game against Aviston Grade School. I scored thirty-one of our thirty-two points, but we still lost the game. While in the seventh and eighth grades, I was also fortunate enough to play on the St. Rose Baseball Team's new entry into the Clinton County Baseball League. It was an excellent semi-pro league with older experienced players, some returning from minor league teams. There were also younger players like me. A picture of the first 1958 St. Rose Entry Team in the Clinton County Baseball League is shown in Figure 76A.

Figure 76A. St. Rose 1960 Initial Entry Team.

In Figure 76A, I am second from left standing, between Vince and Bob Klosterman. Next to Bob Klosterman is their younger brother, Joe. Others standing to the left of Joe are Tom Wilke, Paul Rehkemper, and Ben Wilke. In the front role kneeling from the left, are Marcel Timmerman and Jerry Endres. In the middle is David Woltering, my first cousin and the youngest player on the team. On his left are Junior Jung and Paul Beauregard. While we didn't win many games the first few years, I remember a 1-0 loss in a sixteen inning against a strong Germantown, Illinois team. Toby Haake, the pitcher, had previously pitched in Triple-A minor league but had recently returned to play for Germantown.

I also remember playing on a junior league baseball team in Highland, Illinois, where I often pitched and, in one game, struck out fifteen out of sixteen batters.

I usually had to hitchhike to Highland, Illinois, for the games, about nine miles away. With few cars on the rural highway, I walked much of the time. Occasionally, I would ride my bike. While in high school, I continued to play in the Clinton County Baseball League, but not initially back with the St. Rose team, as other teams recruited me. I played one year for the Aviston, Illinois team and another for the Beckemeyer, Illinois team before returning to play for the St. Rose team. After my junior year at Mater Dei High School, I returned to play with the St. Rose team. As shown in Figure 76B, we had an excellent team, including my two brothers-in-law, Lloyd Castillo and Slugger Lampe, and David Woltering's brother-in-Law, Ernie Hoffman. When major league scouts came in over the 1961 Fourth of July weekend, they signed six players, four being pitchers. As a result, our St. Rose team became a strong contender for the Clinton County Baseball League Championship that year. Several of the six players signed by the major league scouts made the major leagues, e.g., Pat Jarvis with the Atlanta Braves and Tom Timmerman with the Detroit Tigers

76B. Significantly improved St. Rose 1962 Team

HIGH SCHOOL EXPERIENCES AT MATER DEI CATHOLIC HIGH SCHOOL, BREESE, IL

Mater Dei Catholic High School was established in 1959, primarily for Catholics in Clinton County, IL. It consolidated two much smaller Catholic high schools, St. Mary's in Carlyle, IL, and St. Dominic in

Breese, IL. In addition, it opened up an opportunity for other students in and outside of Clinton County. I played on the baseball and basketball teams at Mater Dei High School. A picture of the 1959 Mater Dei High School Baseball Team is shown in Figure 77A. As a freshman, I played on the team and am the second player kneeling from the left. Jerry Peppenhorst, next to me, became my brother-in-law when he married my sister, Juanita. We went on to win the district championship that year.

Figure 77A. 1962 Mater Dei HS BB Team.

A picture of the 1962 Mater Dei High School Basketball Team is in Figure 77A. From the left, I am the third player standing in uniform, with no warmup. I was named the most valuable player for this team. My most significant attribute as a basketball player was my jumping and defense ability, part of it was natural, but jumping with vest and ankle weights in practice also helped. This jumping ability against Marshall HS from Chicago is shown in the paper clipping in Figure 77B. Marshall was one of the best teams in Chicago and won several state championships. Unfortunately, we lost the game by four points but gave an outstanding effort.

Figure 77B. Schrage, 6'2" shoots over and Outjumps Marshall's 6'7" Center

Figure 78A.Student Council President w/ New Class Officers.

My favorite courses in high school were English and Mathematics, although Physics was also of considerable interest. By the time I was an upper-class student, I had become very interested in studying engineering in college. I also became interested in applying for an appointment at the United States Military Academy (USMA) at West Point. Of course, I hoped to combine these interests with an athletic scholarship.

I had become an outstanding athlete in high school and had even been approached by a Detroit Tigers "Bird Dog" Scout, who ran a sporting goods store in Breese, IL, about signing a baseball contract after high school. However, as a 6'2" center in high school, there weren't many major university basketball programs knocking on my door. I applied for an appointment at USMA. I had the grades and leadership experience as the President of the Student Council, as illustrated in Figure 78A, and the most valuable player on the Mater Dei Basketball Team, Figure 78B.

Fig.78B. Basketball MVP w/Coach Killen

However, my congressman only selected me as the First Alternate for USMA. I would only get the appointment if the primary selected applicant didn't pass the required physical exam, which he did. Fortunately, I was offered a combined baseball and basketball scholarship from Quincy College in Quincy, IL and accepted.

Two other of my classmates, Dewey Kalmer and Dennis Trame, teammates with me on both the Mater Dei HS baseball and basketball teams, were also offered baseball and basketball scholarships to Quincy College. Another basketball and baseball teammate, Tom Lager, was offered a partial baseball scholarship. I was excited to have friends and teammates joining me at Quincy College.

COLLEGE AT QUINCY COLLEGE, QUINCY, IL

When I arrived at Quincy College, I selected Mathematics as my major, thinking that I could still possibly transfer into engineering at another school, perhaps USMA or Notre Dame U., one of my top choices. I was very successful in both sports and academics at Quincy College, as I was the only freshman to receive a varsity letter in both basketball and baseball.

Figure 79. Quincy College 1962-63 Varsity Basketball Team, Dan Schrage is 4th from Right, Kneeling.

A picture of the Quincy College Basketball Team is in Figure 79. I found that my jumping ability and quickness were sufficient to succeed in college basketball. I also found that my ability to hit a baseball in college was acceptable, as I was second in home runs on the Quincy College Baseball Team. It was exciting and satisfying that two of my teammates from Mater Dei HS were also successful in both basketball and baseball at Quincy College and beyond. Dewey Kalmer was a very successful point guard and scorer at Quincy College. He went on to play minor league baseball as a catcher, with a "cup of tea" in the major leagues. He later coached baseball at Quincy College and Bradley University and became the Associate Athletic Director at Bradley University. Dennis Trame became a starter on the basketball team and a top pitcher at Quincy College. He, too, had a successful minor league career in baseball. He went on to be a very successful baseball and basketball coach at Mater Dei HS and an outstanding athletic director.

While I enjoyed my stay at Quincy College, I still desired to attend USMA. Therefore, I applied again for USMA. My Quincy College Athletics Director/Basketball Coach and my Baseball Coach supported my application by writing strong letters of recommendation to my congressman and the USMA Basketball and Baseball coaches. Fortunately, I received the appointment to USMA and reported to West Point on July 1, 1963, for what is called "Beast Barracks."

During my last two years in high school, I dated Nancy Granberg from Carlyle, IL. She graduated a year before me and spent two years at Fontbonne College in St. Louis, MO. We became very close and serious, and I hoped we could stay together, and she could support me while I was at USMA. Luckily, she did, and we got married the day after I graduated from USMA, June 8, 1967.

CADET AND ATHLETE AT USMA, WEST POINT, NY

Except for the hazing in *"Beast Barracks,"* I truly enjoyed my time at USMA and graduated on June 7, 1967. I have been proud of this accomplishment and will continue to value it for the rest of my lifetime. My graduation yearbook picture, significant activities/positions, and my roommates' summary of me are shown in Figures 80A and B.

Roommates at USMA usually rotate to new companies after their sophomore (Yearling) year. I was in Company C-2, Second Regiment my first two years, and Company E-4, Fourth Regiment my last two years. In most cases, roommates remain your closest friends for the rest of your lives, especially those during your first two years, especially the Plebe Year, as you are so dependent on each other. One of my Plebe Year roommates, Jerry Walker, remained very close. Jerry ended up being my ranger buddy during Ranger School. Jerry's fiancée, Mary, and my fiancée, Nancy, roomed together the last two years at West Point, first in Highland Falls, NY, and then in Fort Lee, NJ. After we graduated, we

My Roommates Summary

Although a quiet individual, Dan has earned the friendship of everyone he has encountered. He has earned fame and prestige while distinguishing himself as one of Army's finest basketball players. In addition to his talents on the court, Dan also has the potential equally well in academics, but the lure of the more important things of life, sleep in particular, has kept him from wearing stars. No doubt a success in whatever he attempts both in the Army and four years hence. Dan will also carry with him the distinction of being one of West Point's finest. **Activities/Positions:** Class Committee 3,2,1; Class Officer, Athletic Representative; Catholic Choir 4, 3; Baseball 4,3,2, Basketball 4, 3, 3, 1 Team Captain

Figure 80A. My Roommates Summary & USMA Experience Figure 80B. Graduation Photo

Figure 81. My Roommates Last Two Years at West Point

served in each other's weddings and remained close friends afterward. Unfortunately, Jerry passed away from a heart attack in Spring 2021. My roommates for my last two years are shown in Figure 81. Starting from left is Ernie Heimberg, probably the smartest cadet, finishing first in our class. I roomed with Ernie when I was on Brigade Staff. Next to him is Dean Hanson, the starting middle linebacker on the USMA Football Team. Next to me on my right side are Bob Sellars and Jim Warner. Bob was a comedian who kept us in stitches. Jim was probably the most outstanding track and cross-country runner in USMA history. E-4 was known for its athletes and several team captains in basketball, track, and lacrosse.

I was particularly proud of being on the Class Committee, voted in by my classmates, and selected as the Class Athletic Representative. Every cadet at USMA is somewhat of an athlete, or they would not have been selected for USMA. Academically, I finished in the top half of my class, but as indicated in Figure 80A, I didn't focus on academics as much as athletics while at USMA. Fortunately, I was a Math major at Quincy College, so Plebe Mathematics, the most challenging course at USMA, was not as difficult for me.

While I was having much success in sports, i.e., basketball and baseball, and enjoyed boxing and wrestling classes, swimming was an Achille's heel in my sports venue. One of the first things plebes must do in *"Beast Barracks"* is jump in the pool and swim as long as possible. While I learned to swim across the pool and back in Illinois during the summertime, this didn't impress the USMA swimming coaches.

I was therefore assigned to *"The Rock Squad."* This meant that during the academic year, I had to spend one day a week after classes reporting to the USMA swimming pool for swimming lessons. You had to stay on *"The Rock Squad"* until you could demonstrate a capability to pass the swimming test required for Ranger School, four years hence. The swimming test consisted of wearing fatigues and boots with a filled backpack and carrying a rifle. While staying afloat treading water, you had to remove the fatigue pants, tie a knot in the pant legs and establish an air pocket in both legs so that you could at least stay afloat. It took me most of the Fall semester to eventually pass this strenuous test.

However, this wasn't the end of it. During the final semester of my senior, First Class year, before graduating from USMA, those cadets going to Ranger School that had been on the *"The Rock Squad"* had to go back to the USMA swimming pool. We had to verify that we could still complete this test before graduation, which I finally did. Ironically, when I went to Ranger School, initially at Fort Benning, GA, in June 1968, following the Field Artillery Basic Course at Fort Sill, OK, I took the swimming test and was classified as a strong swimmer. This meant that during Ranger School, especially in the Swamp Phase-out of Eglin AFB, I would be carrying the machine gun or radio. These were the heaviest load any ranger candidate had to carry. This bothered me as most of the ranger students were USMA classmates, and I knew they were stronger swimmers than me. Luckily, I never drowned or lost a machine gun or radio when going underwater during the Florida Phase.

My plebe or freshman basketball coach at USMA was a new coach at West Point, named Robert Montgomery Knight, or Bob Knight. He had played on the Ohio State University National Championship Team in 1960. They were a runner-up to the University of Cincinnati in 1960 and 1962, which had Oscar Robertson as their Star. The Head USMA Basketball Coach was Taylor "Tates" Locke, who had played at Ohio Wesleyan. With the military draft in effect in the 1960s, Tates Locke recruited Bob Knight, drafted as a Private in the Army, as his top assistant basketball coach at West Point.

Similarly, the Head USMA Football Coach recruited Bill Parcells as an assistant football coach. It turned out that Bob Knight and Bill Parcells, both Privates, had somewhat similar personalities and became best friends, which continued years beyond USMA when Bob coached at the University of Indiana, and Bill was coaching the New York Giants.

Another part-time assistant coach for Army Basketball who has been respected by every coach and player he assisted over the years was COL(Ret) Tom Rogers, who later became an Assistant Coach at Duke University with Mike Krzyzewski. When I was a plebe, LTC Rogers was a USMA Mathematics Instructor and part-time assistant basketball coach in 1963. Since Bob Knight also was an assistant varsity head coach to Tates Locke, he did much of the scouting of teams USMA was to play, as well as helping with recruiting. When Bob Knight wasn't available to coach the Plebe Team, LTC Tom Rogers served as the Coach. For two of the games and teams we played with LTC Rogers as the coach, I was the highest scorer and rebounder. In one game, I scored thirty-five points and had twenty-one rebounds. In the other game, I had thirty-one points and ten rebounds. I never scored that many points when Coach Knight was the freshman or later head coach. LTC Rogers often told Coach Knight I could score. I ended up being the center on the Plebe Team and the leading scorer for the team, as one of the centers recruited dropped out of USMA during Beast Barracks. It proved to myself and the coaching staff that I would have a role on the USMA Varsity Basketball Team in the following years.

Figure 82A. Silliman Driving vs. Dayton.

The first and only high school All-American basketball player ever to come to USMA to play basketball was Mike Silliman from Louisville, KY. (Figure 82A) His 1,342 points set a career scoring record at West Point, and he missed the last eight games of the 1965-66 Team with injury. He was All-America and All-East his last two years. Mike became the Captain of the 1968 USA Olympics Basketball Team and an NBA player. Mike was also an excellent baseball player, and we often roomed together on USMA baseball trips. I believe the 1964-66 USMA Varsity Basketball Teams had more depth, size, ball handling, and shooting than any previous Army teams. We reached the semi-finals in the National Invitational Tournament (NIT) for three straight years. To find a place and get playing time, I had to concentrate on playing

defense and rebounding, the foundation of Coach Locke and Coach Knight's coaching philosophy. Shown in Figure 82B is Mike Silliman and me battling Sonny Dove from St. John's U. for a rebound. The 1964-65 USMA Team upset highly ranked teams, St. John's and NYU at USMA. In the St. John's Game, I stole the ball in the last minute using our press defense and then made the winning basket in the St. John's Game. St. John's was our biggest rivalry in college basketball during these years. I made the all Eastern Collegiate Athletic

Figure 82B. Schrage & Silliman vs. St. John's Love

Conference (ECAC) Big East Team that week.

Coaches Tates Locke and Bob Knight's philosophy on defense was to play tight man-to-man defense and have one player defend the other team's best player. Beginning in my sophomore (yearling) season, I was most of the time designated defensive player guarding players from 5'8" to 6'8". As a result, I ended up guarding many college basketball All Americans, including Bill Bradley from Princeton (who played in the NBA and went on to be a U.S. Senator). Clyde Lee from Vanderbilt U., Joe Ellis from the University of San Francisco (USF), and Herm Gilliam from Purdue all went on to play in the NBA. Plus, there were several others.

Following my sophomore year, Coach Locke left USMA to take the head coaching job at Miami of Ohio, and Bob Knight became the USMA Basketball head coach. In September 1965, I received a note from Coach Locke wishing me, *"Good luck this season. You are the best defensive player I've ever seen. Stay that way!"*

Also, during the Summer of 1965, Coach Knight sent my parents a note which stated, *"Danny has done an excellent job for us this year. His defensive play is about the best I have ever seen"*. Years later, after being Head Coach at Indiana, Coach Knight reconfirmed in his books and on television that I was the best defensive player he ever coached. This confirmed to me and motivated me to strive to be *Perfect in All Respects* as a defensive basketball player.

Figures 83 A. 1965-66	*Figure 83 B. 1965-66 Team Huddle in St. John's Game*

For the 1965-66 Season, we were very optimistic about being an excellent basketball team that could go deep in the NIT or NCAA Tournaments. Key players from the 1966 Team are illustrated in Figure 83A. From left to right, the team members on the back row are Bill Shutsky, Dick Murray, Jocko Mikula, Dan Schrage, and Bobby Seigle. Mike Silliman, Coach Bob Knight, and Bill Selkie are in the front row. Shown in Figure 83B is the Army Team huddling during a Game with St. John's U. Unfortunately, Mike Silliman was injured in the Rutgers Game in January 1966. He was out during the rest of the 1965-66 Basketball Season, which dampened our hopes for a successful season and an NIT or NCAA bid. However, we won eleven of our last thirteen games. We won the Rutgers Game that Mike was injured in but lost to Canisius to close out January; then won five of the six February games and lost the other by two points on a last-second shot at St. John's. The Navy victory made the record 16-6, same as it was at the same point in 1964, and Army got an NIT bid.

Without Mike Silliman, we had to up our defense and rebounding effort even more as we did against Penn State, a potential NCAA tournament team. We held them to seven points in the first half with a score of 24-7 at halftime. I continued to guard the other team's best offensive player, Carver Clinton.

As summarized in comments to the NIT Press, Coach Bob Knight supported this approach, documented in a New York Post Article, "Li'l Cadet Rises to the Height," after we won the NIT Quarterfinal Team against the USF. Bob Knight's description of me was not rewarding but accurate. *"Ability-wise, most people wouldn't give him a second look as a basketball player, but he fits into our style of play. He's a doggone tough kid, as tough as any kid can be. Though he's only 6'1", he's the best jumper we have had and has out-jumped 6'7" opponents."*

Coach Knight states: *"He's really responsible for a lot of our success in some of our best wins."* Other statements made by Bob Knight were: *"Danny has covered guards, forwards, and post men at various stages of the season."* Knight said. *"He held Carver Clinton of Penn State to one field goal. Danny kept 6'8" Cram of Cornell scoreless for eight minutes. He held Don Freeman of Illinois to one foul shot for 10 minutes, by which time we had a 14 point lead and we beat them by nine, even though Freeman wound up with 31. Schrage also did a fine job against Bruns of Manhattan and Radcliffe of Navy to help us win."*

Our win against USF, one of the favored teams to win the NIT, was as close to perfection as a basketball game can be. It was called incredible by the NIT Press, considering the height and depth we conceded to the 8-point favored USF Dons. Five West Pointers, the tallest of which were Bill Helkie and Ed Jordan, both 6'3", went all but the last forty-five seconds against USF with its towering frontline of 6'8" Erwin Mueller, 6'6" Joe Ellis, and 6'5" Dennis Black.

The New York Post summed it up as follows: *"Army's fantastic runts wrapped all their hustle and muscle around a superb 34 point scoring spree by Bill Helkie with three others, Dan Schrage, Dick Murray, and Bill Shutsky, also in double figures to spring the biggest upset of the 1966 NIT".*

The Cadets walloped USF 80-63 after leading 39-24 at halftime to reach the semi-finals for the third straight year. Figure 84A is a New York Times photo of Bill Shutsky providing a *Perfect Feint*, e.g., faking out the USF front line defendants and passing the ball to Dan Schrage for a

Figure 84 A. Perfect Feint by Shutsky & Dan Scharge tips in Rebound

basket. The USF players in the picture are their front line: Joe Ellis #31, their leading scorer, Dennis Black #15, and Erwin Mueller #51. Also shown is West Point's Bill Helkie, the game's leading scorer with thirty-four points. Dan Schrage is also shown tipping in a rebound. He held Joe Ellis to six points on three field goals, Figure 84B. This win was West Point's biggest win of the season and one of the biggest wins in Army Basketball history.

ARMY (80)	G	F	TP	vs. SAN FRAN (63)	G	F	TP
Jordan	0	8	8	Black	6	2	14
Mikula	0	0	0	Brown	2	1	5
Scnutsky	5	6	16	Ellis	3	0	6
Noonan	0	0	0	Mueller	6	2	14
Helkie	15	4	34	Fortenberry	1	0	2
Hughes	0	0	0	Gumina	3	3	9
Murray	3	5	11	Blum	2	2	6
Isenhour	0	0	0	Wlimore	1	0	2
Schrage	4	3	11	O'Neill	0	0	0
Seigle	0	0	0				
Totals	27	26	80	Totals	24	15	63

Figure 84B. Box Score

In the 1966 NIT Semi-Finals, Army played BYU, an extensive front line. They had two 6'11" rotating centers, Jim Eakins and Craig Raymond, a good small forward in 6'4" Steve Kramer, and two outstanding All-Big West guards, Dick Nemelka and Greg Condon. Army started with the same aggressive defense they had used against USF. BYU failed to sink a shot from the floor in the first seven and a half minutes and was down

14 to 2. Eventually, Eakins came out for Raymond, and BYU worked up to within five points and then a seven-point deficit at halftime. It became clear that even if BYU regained its shooting effectiveness, the game would be decided under the basket. BYU's Steve Kramer, limited to three points in the first half, picked up twenty in the second. However, it was controversial foul shots that would make the difference in the end. As Bob Knight saw it, one of the referees, Eisenstein, was all set to call a violation on Nemelka, but the other referee, Fidgeon, made the call for an Army foul. The foul call came with 2:17 minutes remaining and Army leading 58-56. The foul was the fifth for Army's Bill Helkie, and Dick Nemelka converted both free throws to tie the score for the first time in a one-and-one situation.

Figure 85A. Dan Schrage guarding Nemelka.

A technical foul on Coach Knight resulted in two more points for BYU. Four free throws by Steve Kramer followed this, and BYU won the game 68-60. The controversial foul resulted in a continuing feud between Coach Knight and Referee Eisenstein, which was carried over to the following season and may have kept us out of the NIT. I guarded their high-scoring point guard Dick Nemelka while our front-court guys, Bill Helkie, Bill Shutsky, and Mike Noonan, had to shut down their inside game. The Army had four players in double figures, with sophomore Bill Shutsky leading the way with eighteen points, followed by Noonan with fifteen and Helkie and Murray with twelve each. A photo of Dan Schrage guarding Dick Nemelka is in Figure 85A, with Mike Silliman watching from the Sidelines, Figure 85B. No doubt that if Mike Silliman weren't injured, we would have been NIT Champs in 1966.

Figure 85B. Mike Silliman on Sidelines

In the summer of 1966, I was notified by Coach Knight that I was selected as the Captain of the 1966-67 USMA Varsity Basketball Team. This honor accomplished a primary goal of mine: perfection in athletics. A picture of the 1966-67 USMA Varsity Basketball Team is shown in Figure 86.

Figure 86. The 1966-67 USMA Varsity Basketball Team

I am in the center holding the basketball. As seen in the back row, we had considerable height at the 6'6" and 6'7" levels, although mainly at the sophomore level and inexperienced. Other notables in the picture are Bill Shutsky, our leading scorer and excellent player for his size, kneeling on the left end and Mike Krzyzewski kneeling, an upcoming guard, on the floor, second from the right end. One thing that Bob Knight did was to always list us at shorter heights than we were. For instance, I was listed as 6'2" in high school, while in college, I was more like 6'2.5" or 6'3". I was always listed as 6'1" at USMA.

Figure 87A. Projected Starting Five.

Figure 87A is the initial 1966-67 Team starting five, with Coach Knight. From left to right are Dan Schrage, Billy Shutsky, Mike Noonan, Jocko Mikula, and Eddie Jordan. Illinois: Dan Schrage, Mike Noonan.

Figure 87B. Three Illinois Players

In Figure 87B are three players from and Mike Krzyzewski, Jocko, Eddie, and I were the only remaining players from the 1967 Class. Jocko was an excellent all-state high school player from Dayton, OH. Ultimately, he became one of the Army's best shooters and defensive players. Eddie was from Montgomery, AL, and became the best scorer in our class during his last two years at USMA. We bonded together during our senior (first-class) year, as we faced a demanding schedule.

An article out of New York entitled "Army Cagers Face Tough Schedule" stated: *"The Army basketball team under the leadership of Coach Bob Knight and Team Captain, Dan Schrage, will face the toughest schedule in Army basketball history as the Knights hope to better the record it posted last year. Once again, the defense will be the key to victory, especially against such teams as Ohio State, Princeton, Purdue, St. John's, and NYU. The schedule includes only nine home games with twelve away games. Coach Knight is looking forward to the game with Ohio State, still coached by Fred Taylor, Knight's playing mentor during his playing days at Columbus."*

My assessment of the 1966-67 USMA Varsity Schedule in an interview for Slum and Gravy magazine was that it would be more challenging than last year. We played most of our games on the road, and most were against outstanding opponents. Some of the better teams I expected to be tough were Ohio State, Princeton, Syracuse, St. Johns, NYU, Purdue, and the teams we would play at the Davidson Tournament at Charlotte, NC.

Another factor that would make this year a tough one was that no one would be taking us for granted, not after the success we had the past two or three years. We very much wanted to play in the NIT again, but we also knew that there would be a lot of schools gunning for us.

We started with three losses against Princeton, Syracuse, and Cornell. Coach Knight didn't take losses well, and when we returned from Cornell, we went straight to the Field House to practice, even though it was past midnight.

We then followed with two wins against Lehigh and Holy Cross at home. We then went on the road over Christmas Break and played at Purdue, where we lost in double overtime on December 21, 1966, where I guarded Herm Gilliam, who went on to play in the NBA. We followed this up on December 23, 1966, with a game at Ohio State, which we lost by two points when their leading scorer made a 22-foot jump shot at the buzzer.

This was a significant loss for Bob Knight. Usually, after a game, we would wait in the dressing room until Bob Knight came in and gave us a debrief on our performance before we would take our showers. After this game, he did not come for about an hour; however, I did not let anyone take a shower until he came. After about an hour, Coach Knight came in with Coach Fred Taylor. You could see that Coach Knight took this loss extremely hard. Coach Taylor gave us a debrief and told us we played a great game, especially on defense, holding his Team to sixty-one points.

Following this game, we took a quick Christmas Break until we played Fordham U. and the University of Maryland in the Charlotte Invitational Tournament on December 28-29, 1966. We were hoping to win the first tournament that Army Basketball had ever won. We beat Fordham but lost the Maryland game, 57-53, one we should have won but didn't. This was the make-or-break point for our season. Coach Knight emphasized us three seniors, Jocko, Eddie, and I, to get the team on track. Starting with a Seton Hall win on January 7, 1967, we won the next nine out of eleven games with only the Navy Game remaining. We had never lost to the Navy in our four years at West Point, and we didn't lose this one, finishing with a 13-8 record. We were told informally that we would probably get an NIT Invitation if we beat the Navy. We beat the Navy by ten points but didn't get an NIT Invitation. There was some speculation that it was due to the Coach Knight and Referee Eisenstein encounter during the BYU Game in the 1966 NIT. This was the most demanding season I ever had in sports, but I felt we had left a winning legacy for USMA Basketball.

Coach Bob Knight didn't like to lose, and it took a lot out of us three seniors to right the ship. We learned a lot but also witnessed the tenacity of Coach Knight and what it took to play for him. While being a tough kid, I never doubted Coach Knight's motives and have expressed my loyalty to him, as he has to me over the years. However, a lot of kids don't have that toughness. During my first three years at West Point, I played baseball and was the leading hitter during my first year, e.g., Plebe Year. During my second year, Yearling Year, Mike Silliman and I came out late for the Baseball Team, following participation in the NIT Tournament. On inclement days the USMA Baseball Team had workouts in the old Field House.

During my first workout in the Field House, after hitting and running out to first base on the last hit ball, I pulled my hamstring quite severely and had to drop out of baseball for the rest of Spring 1965. I came out again in Spring 1966 and played considerably and received a varsity letter on the USMA Varsity Baseball Team. The Team is shown in Figure 88. Mike Silliman and I often roomed together on baseball trips. He was an excellent baseball player and golfer, and an All-American basketball player.

Figure 88. 1966 USMA Baseball Team with Dan Schrage & Mike Silliman named

However, it was hard to keep up with him when drinking beer. After your second or third year at USMA, most cadets went on summer training with Army units, some in Europe and some in the States. Mike went to Germany and enjoyed the beer. When he came back in the Fall of his junior (second class year), he was about 20-30 pounds heavier. Coaches Locke and Knight got him on a diet and an extreme exercise program before the 1965 USMA Basketball Season. For the Spring of my senior 1967 First Class Year, I opted out of playing baseball to have a nose operation to repair a deviated septum. This was necessary for me to be eligible to attend flight school and become an Army Aviator after at least a year of service in a combat arms branch. This ended my college baseball career, which was disappointing because, in many ways, I thought that baseball was my best sport.

My third class (Yearling), second class (Cow), and first class (Firstie) years were very enjoyable. During the Yearling Summer, 4-Week Cadet Field Training (CFT) was conducted at Camp Buckner. In the evenings, we would sometimes play pick-up basketball games with Coach Knight and his former teammate at Ohio State, John Havlicek, occasionally watching.

During the second-class Cow Summer, shortly after the academic year, we went on a tour of U.S. Army bases covering the different branches of the Army, i.e., Armor at Fort Knox, KY; Field Artillery at Fort Sill, OK; Air Defense at Fort Bliss, TX; Infantry at Fort Benning, GA; and

Engineer at Fort Belvoir, VA. This was to determine which Branch of the Army we would select. My initial preference was for Air Defense Artillery or Field Artillery, although the choice would depend on my academic class rank.

This was followed by a month off at my home in Illinois in July 1965. My parents and I took a one-week road trip to Colorado. It was extremely hot crossing Kansas, but the mountains in Colorado were a welcome relief. In August, I returned to West Point to be on the Second-Class Detail at Camp Buckner.

During my next year's Firstie Summer, I spent four weeks at Fort Hood, Texas, with the 1st Armored Division. I served as a Third Lieutenant platoon leader of an Infantry Platoon. This was not a pleasant experience, trying to keep up with the tanks in our tin can personnel carrier vehicles, as well as camping out at night and trying to avoid tarantulas and scorpions.

At the beginning of the Firstie academic year, a Ring Ball was held for graduating cadets to receive their class rings. Shown in Figure 89 A are my fiancée Nancy Granberg and I at the Ring Hop. It was a great night. Another picture of me in my seldom-worn Class White Uniform is in Figure 89B.

Figure 89A. Dan and Nancy at the Ring Ball

Figure 89B. Dan in his Class A Uniform

Graduation Day on June 7, 1967, came quickly. Shown in Figure 90A is Cadet Daniel P. Schrage receiving The Eber Simpson Memorial Trophy presented to the captain of the 1967 Basketball Team.

Figure 90A. Cadet Schrage Eber Simpson Trophy.

Figure 90B. Cadet Schrage Diploma from MG Bennett

Figure 90B is MG D.V. Bennett, USMA Superintendent, giving the Bachelor of Science Diploma to Cadet Daniel P. Schrage. Later, 1LT Daniel P. Schrage would cross paths with LTG Bennett again when he was the VII Corps Commander in Europe. This included visiting 1LT Schrage Honest John Missile Battery in Germany at a field location during the Return of Forces to Germany (REFORGER) I Exercise in February 1969.

JUNE 8, 1967, WEDDING BETWEEN 2LT DANIEL P. SCHRAGE AND NANCY GRANBERG

Nancy and I had dated each other for seven years; two years in high school, one year while I was at Quincy College, and Nancy was at Fontbonne College, plus four years while I was at West Point. We got married on June 8, 1967, at the West Point Catholic Chapel. Coach Bob Knight and his wife, Nancy, were in attendance.

In Figure 91A, Nancy and Dan exit the Chapel under swords from Dan's Classmates, a West Point tradition. It includes sealing with a kiss in Figure 91B.

The entire wedding party is shown in Figure 92A, with Dan's Parents in Figure 92B. Dan's Cadet friends, Jerry Walker and Jocko Mikula were groomsmen in the wedding party, with Nancy's brother, Steve, as the best man. The bridesmaids were Nancy's cousin, Mary, and Mary Walker, with Jan Powell as the maid of honor. In addition, Nancy's young cousins were included as flower girls and as junior groomsmen.

Figure 91A. Dan and Nancy are exiting under swords.

Figure 91B. Sealed with a Kiss.

Figure 90A. Entire Wedding Party.

Figure 92B. Dan's parents, Al and Mary Schrage.

Following the wedding, Dan and Nancy drove to Montreal, Canada, for their honeymoon. They took in the Montreal Expo 67, Canada's World Fair. Upon returning from their honeymoon to Illinois, Dan and Nancy moved to Lawton, OK, for Dan to attend the Field Artillery Basic Course before attending Ranger School. Dan had selected Field Artillery as his combat arm, intending to go to flight school following a tour in Germany. These assignments led to many moves, including in the USA and Germany.

Chapter Four

Experiences in Military Courses and as a Nuclear Weapons Field Artillery Battery Commander in Germany

- 1.Experiences in Field Artillery Basic Course and Ranger School
- 2.Experience as a nuclear weapon Battery Commander in West Germany, 1968-1969
- 3.Experiences from Rotary Wing Flight School and Deployment to South Vietnam

Experiences in Field Artillery Basic Course and Ranger School

Field Artillery Basic Course

In June 1965, cadets moving into their second class (junior) year took their annual branch trip to visit different Army posts to get an understanding of the branches they may want to choose in their first-class (senior) year. Each Army Branch makes a big pitch to recruit cadets to their branch.

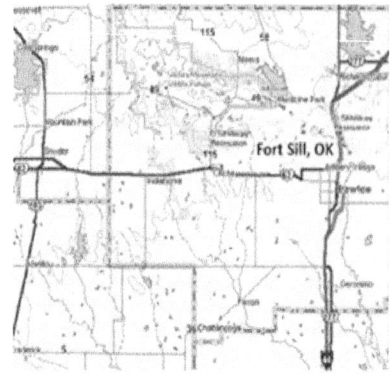

Figure 93 is a Map of the Fort Sill, OK, Area.

I chose Field Artillery and Missiles Branch at Fort Sill, OK, as my primary branch for several reasons. First, Fort Sill had been the first Army Aviation Center and was developing aerial rocket artillery, compatible with my desire to be an Army Aviator. Second, in addition to tube artillery, they had rocket and missile units, which I preferred. Third, I enjoyed Fort Sill's location next to the Wichita Mountains' large Wildlife Refuge with thousands of buffalo, which we saw. We also enjoyed a barbecue during our branch trip. Figure 93 is a map of the Fort Sill Area, the Wichita Mountains Wildlife Refuge, and Fort Sill Military Reservation. Shown in Figure 94 are the Fort Sill examples of a Missile and Howitzer. While the Field Artillery Basic Course was only six weeks, the Advanced Course would be nine months, followed by my year in South Vietnam in Spring 1971.

Figure 94. Fort Sill. Howitzer and Missile/Rocket

History Summary of Fort Sill, OK Area

The War Department combined all artillery training and development under the U.S. Army Artillery Center at Fort Sill in 1946. The center included the Artillery School, the Antiaircraft and Guided Missile School at Fort Bliss, Texas, and the Coast Artillery School at Fort Scott, Calif. The air defense artillery became its own branch in 1966. In 1953, school personnel fired the first nuclear-capable field artillery gun (the 280mm gun known as Atomic Annie) at Frenchman's Flat, Nevada.

During the 1950s, school personnel also helped develop rocket and missile warfare (the U.S. arsenal included the Honest John rocket, Corporal missile, and Redstone missile) that could carry a nuclear warhead. In 1963, the school tested aerial rocket artillery, which equipped helicopters with rockets. As was demonstrated in the Vietnam War, aerial rocket artillery was very effective.

The school cooperated in developing the Field Artillery Digital Automated Computer, commonly called FADAC, in computing fire direction data. Introduced in 1966-67, FADAC made the field artillery a leader in computer development for the Army. After the Vietnam War, the school introduced the Multiple-Launch Rocket System, the Army Tactical Missile System, the Paladin 155-mm, self-propelled howitzer, and other field artillery systems. However, the Field Artillery School's interest in aerial rocket artillery waned, which was picked by the Armor School. The Infantry School picked up utility helicopters responsibility.

Ranger School

History and Background

The United States Army Ranger School is an approximately 60-day small unit tactic and leadership course for developing functional skills directly related to units whose mission is to engage the enemy in close combat and direct-fire battles. Ranger training was established in September 1950 at Fort Benning, Georgia, and has changed little since its inception. Until recently, it was an eight-week course divided into three phases. The course is now 61 days and is divided into three phases: the Benning Phase, the Mountain Phase, and the Swamp Phase. Ranger School is open for soldiers, marines, sailors, and airmen in the U.S. Armed Forces and select allied military students and is conducted at various locations. The Benning Phase occurs in and around Camp Rogers and Camp Darby at Fort Benning, Georgia. The Mountain Phase is completed at Camp Merrill, in the remote mountains near Dahlonega, Georgia, and the Swamp Phase is conducted in the coastal swamps near Camp Rudder, Eglin Air Force Base, Florida. In 1966, a panel headed by General Ralph E. Haines Jr. recommended making Ranger training mandatory for all Regular Army officers upon commissioning. On 16 August 1966, the Chief of Staff of the Army, General Harold K. Johnson, directed it, and this policy was implemented in July 1967. It was rescinded on 21 June 1972 by General William Westmoreland. Once again, Ranger training was voluntary.

According to U.S. Army regulations, those graduating from Ranger School are presented with the Ranger Tab, Figure 95A, which is worn on the upper shoulder of the left sleeve of a military uniform, as illustrated in Figure 95B. Wearing the tab is permitted for the remainder of a soldier's military career. The cloth version of the tab is worn on the Army Combat Uniform and Army Green Service Uniform; a smaller, metal version is worn on the Army Service Uniform.

Figure 95A Ranger Tab

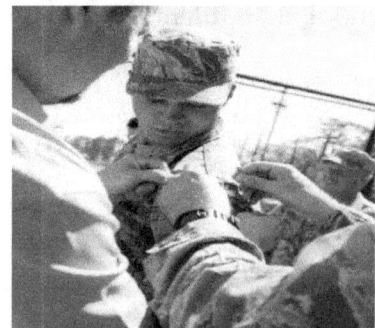

Figure 95B: Ranger Tab on Upper Shoulder

MY STORIES FROM RANGER SCHOOL

The policy requiring combat arms officers to go to Ranger School was implemented in July 1967. Thus, my USMA graduating class was the first class that was required to attend Ranger School, although many prior USMA graduates volunteered to go to Ranger School. At first, I was disappointed that Ranger School was required for USMA graduates going into a combat arm in the Regular Army. However, having gone through the Ranger School Combat Water Survival Test at USMA and being in the best shape of my life, I felt ready. In addition, I believe Ranger School was valuable for me in teaching teamwork and later combat assignments.

As stated in Chapter Three, Jerry Walker was my Ranger Buddy. We arrived at Camp Darby, Fort Benning, GA, in late September 1967 for the Benning Phase. This was followed by the Mountain and the Swamp Phases, with graduation scheduled at the end of November 1967. This meant that the Mountain Phase would be in late October and early November 1967, when it could likely be below freezing. My additional stories will follow the description of each phase. (Army Ranger School Is a Laboratory of Human Endurance Apr 20, 2020)

The "Benning Phase"is the "crawl" phase of Ranger School, where students learn the fundamentals of squad-level mission planning. It is "designed to assess a Soldier's physical stamina, mental toughness, leadership abilities, and establishes the tactical fundamentals required for follow-on phases of Ranger School". In this phase, training is separated into two parts, the Ranger Assessment Phase (RAP)and Squad Combat Operations. The Ranger Assessment Phase is conducted at Camp Rogers. Ot emcp,[asses Daus 1-3 of training. Historically, it accounts for 60% of students who fail to graduated Ranger School. In Figure 5 MAJ Jaster, first femail Army Reserve Officer to graduate from Ranger School in October 2015

Figure 94. Instructor performs a fireman's carry-on, a simulated casualty during the first phase of Ranger School

BENNING PHASE

RANGER PHYSICAL FITNESS TEST (RPFT) REQUIRES THE FOLLOWING MINIMUMS:

- Pushups: 49 (in 2 minutes, graded strictly for perfect form)
- Chin-ups: 6 (performed from a dead hang with no lower body movement)
- 5-mile individual run in 40 minutes or less over a course with gently rolling terrain
- Combat Water Survival Test (Test I prepared for at USMA but no longer conducted as of 2010)

A student's graduation is highly dependent on their performance in graded leadership positions. This leadership ability is evaluated at various levels in various situations and is observed in one of typically two graded leadership roles per phase. The student can either meet the high standards and be given a "GO" by the Ranger Instructors (RI) or can fail to meet this standard and receive a "NO GO." The student must demonstrate the ability to meet the standard to move forward and can thus only afford one unsuccessful patrol.

MY STORY

The RPFT was tough. As mentioned earlier, I passed the Combat Water Survival Test, thanks to the time and effort I was forced to put in on it while at USMA. There was a lot of pressure in Ranger School to pass graded leadership roles as patrol leader as early as possible. In 1967 there were six leadership roles as patrol leader, usually two per phase. One had to pass over half of the leadership roles to get a "GO" and earn a Ranger Tab.

Often time, the failure may not be the patrol leader's fault. I remember one instance where one of the Ranger students in our Long Range Reconnaissance Patrol (LRRP) Team left his rifle at an earlier rest stop without telling the patrol leader being graded. The patrol leader was given a "NO GO."

I was fortunate to receive a "GO" on my first three patrol leader opportunities before the end of the Mountain Phase, which meant I only needed one more "GO" to stay in Ranger School and be eligible for my Ranger Tab. My Ranger Buddy, Jerry Walker, and I were a great team and successfully got our Ranger Tabs.

MOUNTAIN PHASE

(Army Ranger School Is a Laboratory of Human Endurance Apr 20, 2020) Ranger School Is a Laboratory of Human Endurance Apr 20, 2020)

The second phase of Ranger School is conducted at the remote Camp Merrill near Dahlonega, GA, by the 5th Ranger Training Battalion. According to the Ranger School, here "students receive instruction on military mountaineering tasks, mobility training, and techniques for employing a platoon for continuous combat patrol operations in a mountainous environment."

Adding to the physical hardships endured in the Benning phase, in this phase," the stamina and commitment of the Ranger students are stressed to the maximum. At any time, they may be selected to lead tired, hungry, and physically expended students to accomplish yet another combat patrol mission."

The Ranger students continue to learn how to sustain themselves and their subordinates in the mountains. The rugged terrain, severe weather, hunger, mental and physical fatigue, and the student's psychological stress allow them to measure their capabilities and limitations and those of their fellow soldiers. In addition to combat operations, the student receives four days of military mountaineering training. The training ends in a two-day upper mountaineering exercise at Yonah Mountain, GA, to apply the skills learned during lower mountaineering. Each student must make all prescribed climbs at Mt. Yonah to continue the Ranger Course.

A Ranger Instructor (RI) explains the technical aspects of rappelling in Figure 97A. A student receives instructions on rappelling in Figure 95B. One of the Ranger students in our class slipped on a wet cliff and broke his back.

Figure 97A. Technical Aspects of Rappelling.

Figure 97B. Student Rappelling Instructions

During the field training exercise (FTX), students execute a mission requiring mountaineering skills. Combat missions are against a conventionally equipped threat force in a mid-intensity conflict. These missions are both day and night in a two-part, four, and five-day FTX. They include moving cross country over mountains, vehicle ambushes, raiding communications and mortar sites, river crossing, and scaling steeply sloped mountainous terrain.

MY STORY

In the Fort Benning, Mountain, and Swamp phases, Ranger candidates were formed into six-man LRRP teams. Most of the time, these LRRP teams operated independently. However, at times in the Mountain and Swamp phases, they were brought together as five or six teams for larger-scale operations.

In 1967 Mountain Phase Ranger LRRP teams reached their objective in several ways: cross-country movement and a 10-mile FTX march near the Tennessee Valley Divide in the North Georgia Mountains. The

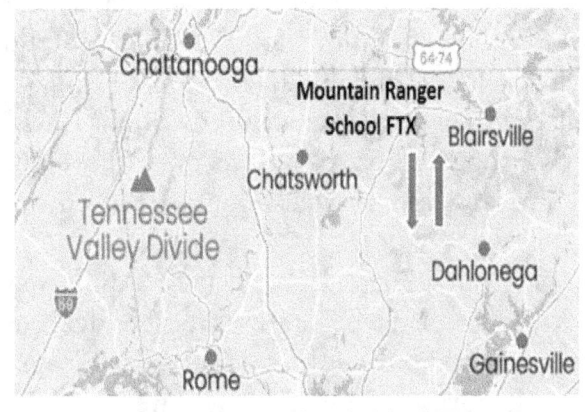

Figure 98. Mountain Ranger School FTX

Tennessee Valley Divide is the boundary of the drainage basin of the Tennessee River and its tributaries.

Illustrated in Figure 98 is the location of the Mountain Ranger School FTX. The arrows show notional truck routes that position and retrieve the Ranger LRRPs for the FTX. The Tennessee Valley Divide coincides with the (https://en.wikipedia.org/wiki/Eastern_Continental_Divide) Eastern Continental Divide to another triple point in northern "https://en.wikipedia.org/wiki/Georgia_(U.S._state)"Georgia. From this point, streams flow northward toward the Tennessee River, eastward to Georgia's Atlantic coast, and southward toward the Tennessee Valley.

By this time, famine was setting in as the LRRP Teams had been on the move without meals for a few days. I remember after being out for a few days, our LRRP was thrown a live chicken by our RI, and we had to figure out how the six of us could make the most out of it for supper. We broiled the chicken and made a chicken broth for sharing. It was the best way between six hungry Ranger students. Before we started the 10-mile FTX march across a part of the Tennessee Valley Divide to attack an aggressor headquarters, we were organized into five LRRPs of six rangers each, to plan our attack across the Divide. We then broke up into our individual LRRP to attack different sectors along different routes. However, there was intelligence in our LRRP from members from an earlier Ranger class that a farmhouse across the Divide had food to sell. We then proceeded to the lights we saw on the other side of the Divide, found the farmhouse, and purchased some food. This raised our spirits and helped temporarily satisfy our hunger; however, we would have gotten kicked out of Ranger School if word got out to the RIs.

Another story from the Mountain Phase was when we went out on an LRRP where we had to cross a small, raging river to attack our enemy objective. It was frigid, with the temperature in the teens. To cross the river without getting our clothes and weapons wet, we had two Ranger candidates who were strong swimmers (not me) in our LRRP who swam across the river naked, carrying a cable that could be tied to a tree across the river. This cable was pulled tight enough so that the rest of our LRRP Team could get naked and carry our backpacks and weapons above our heads. We then proceeded and successfully attacked and achieved our enemy objective.

However, on our way back, our Ranger RIs told us that time was critical for us to meet the trucks that would be picking us up at a rendezvous site. If we missed the trucks, we would have to force march back to our Ranger base camp, over ten miles away. Therefore, when we got to the raging river coming back, we didn't repeat the earlier procedures; instead, each Ranger candidate crossed the river individually, fully clothed, with their equipment. Once across the river, we ran the rest of the way and just made it to the truck rendezvous site in time.

However, with the temperatures in the teens, a great concern of our RIs and us were about frostbite. This was especially critical for the two Navy Seals in our LRRP, as they both had previous frostbite cases. Our RIs were even more concerned. So once we were on the trucks, the RIs would stop them every few miles, have us get out of the trucks, and do calisthenics to keep our blood flowing to prevent frostbite. I had tingling in my feet and thought I had frostbite, but I did not. However, I believe several others did.

SWAMP PHASE

The third phase of Ranger School is conducted at Camp James E. Rudder (Auxiliary Field #6), Eglin Air Force Base, Florida, and taught by the 6th Ranger Training Battalion. Figure 99A shows the Welcome Sign to Camp Rudder and an aerial view in Figure 99B, including the Reptile House.

Figure 99A. Camp Rudder Welcome Sign.   *Figure 99B. Aerial View of Camp Rudder*

According to the Ranger Training Brigade, this phase focuses on the continued development of the Ranger student's combat arms functional skills. Students receive instruction on waterborne operations, small boat movements, and stream crossings upon arrival. Practical exercises in extended platoon-level operations, executed in a coastal swamp environment, test the students' ability to operate effectively under extreme mental and physical stress conditions. This training further develops the students' ability to plan and lead small units during independent and coordinated airborne, air assault, small boat, and dismounted combat patrol operations in a low-intensity combat environment against a well-trained, sophisticated enemy.

The Swamp Phase continues the progressive, realistic OPFOR (opposing forces) scenario. As the scenario develops, the students receive "in-country" technique training that assists them in accomplishing the tactical missions later in the phase. Technique training includes small boat operations, expedient stream crossing techniques, and skills needed to survive and operate in a rainforest/swamp environment by learning how to deal with reptiles and determine the difference between venomous and venomous non-venomous snakes. Camp Rudder has specially trained reptile experts who teach the students not to fear the wildlife they encounter.

Today, Ranger students are updated on the scenario that eventually commits the unit to combat during techniques training. According to the Ranger School Training Phase, the 10-day FTX comprises "fast-paced, highly stressful, challenging exercises in which the students are evaluated on their ability to apply small unit tactics and techniques during the execution of raids, ambushes, movements to contact, and urban assaults to accomplish their assigned missions." The 1967 capstone of the course was an extensively planned raid of the European Liberation Front's (ALF) island stronghold. It required crossing a small bay in the Gulf of Mexico, Figure 100A. This required a small boat operation, illustrated in Figure 100B.

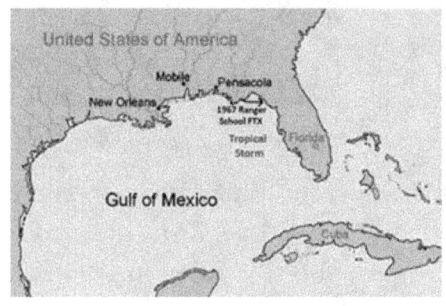

Figure 100A. 1967 Ranger FTX in the Gulf of Mexico.

Figure 100B. Typical Launch-off by LRRPs.

It involved each platoon in the class, all working together on separate missions to take down the simulated cartel's final point of strength. Afterward, students who have met graduation requirements spend several days cleaning their weapons and equipment.

The first Ranger School Graduation in 1950 at Fort Benning, GA, is shown in Figure 101A. The 1967 Ranger Graduation took place at Hurlburt Field., Figure 101B.

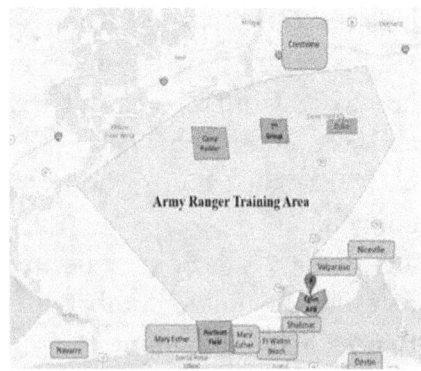

Figure 99A.1950 Ranger School Graduation

Figure 99B. Army Ranger Training Area w/ Camp Rudder and Hurlburt Field

MY STORY

The Swamp Phase consisted of several short exercises, usually several days out and one day back in Camp, and one ten-day capstone FTX Operation. On one exercise, I remember there was an ambush by the aggressors while we were being transported back to base camp by trucks. Since we were slow to respond to the ambush, we were sent back out for another few days' exercise. The most extensive operation was a capstone FTX exercise to attack an aggressor camp across a Gulf of Mexico bay in combat rubber raiding crafts, as illustrated by the arrow in Figure 100A. It was to be a nighttime operation including LRRPs from each platoon in a separate combat rubber raiding craft. There was concern by the RIs about a tropical storm forming off the coast, also illustrated in Figure 100A. A sample departure is in Figure 100B.

Figure 102A. FTX LRRPs are launched. *Figure 102B. FTX LRPs Separated*

Shown in Figure 102A are typical launched FTX LRRPs. It wasn't long before it became dark, and the tropical storm hit with its high winds, causing rough seas, which soon separated the LRRP boats from each other, as shown notionally in Figure 100B. After about an hour at sea, we could not see any of the other LRRP boats as the storm kept blowing. I remember other Ranger students and I got cramps in our legs and had to rotate positions as best we could.

We continued through the rough sea, and after another hour or two, we saw some lights on the distant shore. We headed for the lights. We landed and approached the house with lights. When we got there, we knocked on the door, and a woman with three scared small children behind her opened the door. Our haggard appearance didn't scare her, and we explained our situation and hunger symptoms. The date was a few days before Thanksgiving, 1967, and she offered us food from their Thanksgiving Dinner she was preparing for her husband, who was out of town but coming back for Thanksgiving. We readily accepted and told her we would send her money as soon as we got back to Camp Rudder, which we did.

Figure 103. Sample Picture of Conditions after Tropical Storm Passed.

When we left her house for the beach, the tropical storm had passed, as illustrated notionally in Figure 103. When we walked up the beach, we ran into a Coast Guard boat looking for lost Ranger students like ourselves. Our secret dinner had to be kept hidden from the other Ranger students and the RIs. Needless to say, our morale went up 100%.

Extreme hunger, weight loss, and fatigue were significant outcomes of our Ranger School experiences. However, it was not uncommon, as described in the following sections, the graduation rates and a summary of the physical effects of Ranger School experiences.

(Incidences of Illness and Injury in Army Ranger School, posted June 10, 2015)

GRADUATION RATES

Historically, the graduation rate from Ranger School has been around 50%, but this has fluctuated. Before 1980, the Ranger School attrition rate was over 65%. 64% of the Ranger School class graduated. The graduation rate has dropped below 50% in recent years: 52% in 2005, 54% in 2006, 56% in 2007, 49% in 2008, 46% in 2009, 43% in 2010, and 42% in 2011. Recycles are included in the graduation rates. Recycles are those candidates who begin the program again and are tracked by the class with which they start and affect only the graduation rate of that class.

Physical effects from Ranger School

Following the completion of Ranger School, a student will usually find himself "in the worst shape of his life." As mentioned before, military folk wisdom has it that Ranger School's physical toll is like years of natural aging. High levels of fight-flight stress hormones (epinephrine, norepinephrine, cortisol), standard sleep deprivation, and continual physical strain inhibit full physical and mental recovery throughout the course. Common maladies during the course include weight loss, dehydration, trench foot, heatstroke, frostbite, chilblains, fractures, tissue tears (ligaments, tendons, muscles), swollen hands, feet, knees, nerve damage, loss of limb sensitivity, cellulitis, contact dermatitis, cuts, and insect, spider, bee, and wildlife bites. Because of the physical and psychological effect of low-calorie intake over an extended period, it is not uncommon for many Ranger School graduates to encounter weight problems as they return to their units and their bodies and minds slowly adjust to routine again.

A drastically lowered metabolic rate, combined with a nearly insatiable appetite (the result of food deprivation and the ensuing survivalist mentality), can cause quick weight gain, as the body is already in energy (fat) storing mode.

Food and Sleep Deprivation

A Ranger student's diet and sleep are strictly controlled by the RIs. During the garrison, students are given one to three meals a day but forced to eat extremely quickly and without any talking. During field exercises, Ranger students are given two MREs (Meal, Ready-to-eat) per day but are not allowed to eat them until given permission. This is enforced most harshly in the Darby and Mountain phases. Since food and sleep are at the bottom of the priorities of those in the infantry behind security, weapons maintenance, and personal hygiene, it is generally the last thing Ranger students are allowed to do. As such, the two MREs usually are eaten within three hours of each other, one post-mission and the other before the planning portion of the mission. Though the Ranger student's daily caloric intake of 2200 calories would be more than enough for the average person, Ranger students are under such physical stress that this amount is insufficient. The Ranger Training Brigade does not maintain weight information in the 21st century, but in the 1980s, Ranger students lost an average of 25–30 pounds during the Ranger course.

MY STORIES

My Ranger Buddy, Jerry Walker, and I graduated in late November 1967 at Hurlburt Field, FL. While we were standing in the flight line waiting for the ceremony to begin, a pigmy rattlesnake was slithering through our ranks. Bobby Franks, a graduating Ranger student and USMA classmate from Marianna, FL, reached down and picked up the snake which bit him on his hand. Bobby was experienced with pigmy rattlesnakes. However, the snake bite took him to the hospital and out of the graduation ceremony.

My wife, Nancy, drove our car down from Illinois for the graduation. We planned to drive to New Orleans, LA, and ship the car to Germany. We were assigned to the 1st Battalion of the 34th Artillery, an Honest John Missile Battalion with the 24th Infantry Division in Munich, Germany. While we made it to New Orleans and shipped the car, my weight loss, craving for sweets, and sleepless nights finally caught up with me. When I entered Ranger School, I weighed approximately 195 pounds. However, when I graduated, I weighed about 165 pounds. This was a 30-pound weight loss, typical of the statistics identified above.

Another problem I encountered was the craving for sweets. It didn't help that we were arriving in New Orleans during their Pecan Praline Festival. I ate so many pecan pralines that my digestive system couldn't handle them, and I had to make numerous trips to the bathroom, both during the day and night.

Fortunately, we made it back to Illinois and away from the pecan pralines. We had a great few weeks, including Christmas and New Year's in Illinois, with both of our families. We were excited about our new assignment to Germany. Once I recovered from the physical effects of Ranger School, I reflected on what I had learned. It later turned out to be of great benefit, both in the winter exercises in Germany and in the following assignment to South Vietnam, where there would be a possibility of becoming a prisoner of war (POW) in the mangrove swamps of the Mekong Delta.

EXPERIENCE AS A NUCLEAR WEAPON BATTERY COMMANDER IN WEST GERMANY

Nancy and I arrived in Munich, Germany, in early January 1968. We checked in at the 24th Infantry Division Artillery Headquarters and our assignment to the 1st Battalion of 34th Artillery (HJ) HQs in Will Kaserne, Figure 104A. We were assigned to housing in Warner Kaserne, Figure 104B.

Figure 104A. 24th Infantry Div. Artillery HQs, Will Kaserne.

Figure 104B. Warner Kaserne Housing Area

Shown on a North Munich (München) topological map, Figure 105, are Warner Kaserne, Legend 1, Will Kaserne, Legend 3, Oberschleissheim Army Airfield, Legend 2, and the NATO Nuclear Weapons Storage Area, Legend 4. The " Road to Dachau " is also illustrated on the top of the map. Dachau was one of the major Nazi concentration camps during World War II. It was also the location of a maintenance facility for the 24th Infantry Division in 1968, which we used when I was a maintenance officer with B Battery, 1stBattalion of the 34th Artillery, before becoming the B Battery Commander. This usage will be discussed in later sections.

Figure 105. Topological Map of North Munich, Germany, with key-labeled locations

A picture of an Honest John Rocket/Missile System (HJ) on its 5-ton launch vehicle is shown with its crew in Figure 106A. Firing an HJ in a field exercise is shown in Figure 106B.

Figure 106A. Honest John on Launch Vehicle.   *Figure 106B. Honest John Firing*

The main difference between a rocket and a missile is that missiles have a guidance system to meet their designated target. Rockets, however, do not have a guidance system and are launched in the direction of the primary target using a timing device and wind measurement.

However, HJs with tactical nuclear warheads had a simplified guidance system. The first ballistic missile was the V-2 rocket created in Nazi Germany during World War II. Invented by Walter Dornberger and Werner von Braun, it was first used in 1944 to attack London, England. Following World War II, Werner von Braun brought his team to the U.S.

The elements that make up an Honest John system are illustrated in Figure 107A. Setting the timing for an Honest John (HJ) Missile/Rocket in the field is shown in Figure 107B.

It wasn't long after we arrived that our battalion was getting ready for its annual two-week FTX training and qualification exercise at the Grafenwöhr Training Center (GTC). It is shown on the map in Figure 108A and the Hohenfels Training Center (HTC) in Figure 108B. It shows the Artillery Impact Area at the GTC, where we had to fire our HJ training rockets into an area about the size of a football field.

It wasn't long before West Germany became covered in snow. Shown in Figure 109A is Nancy standing in the snow outside our Warner Kaserne housing. Figure 109B from the left are second Lieutenants (2LTs) Jan Askman, Dan Schrage, and Rick Hill, plus another officer from the 1st Battalion, 34th Artillery, in the snow at GTC in late January 1968. We would be out in the field for three days, followed by a day at our basecamp, which was in the town of Vilseck, Germany.

MY STORIES

Munich was a beautiful city, close to the Alps, and the living quarters at Warner Kaserne were good. With their wives, two of my USMA classmates, Jan Askman and Rick Hill, were also assigned to the 1st Battalion, 34th Artillery. We were all welcomed by LTC Pfeifer, CO, and the other officers in the Battalion. Shortly after we signed in, we left with another couple on a train from Munich to the German port city of Bremerhaven to pick up our cars and drive them back to Munich. It wasn't long before the battalion left for the GTC for its annual FTX training and qualification. While the men were gone, several wives being lonely, visited the Hundemark (dog market) in downtown Munich and purchased pet dogs. Germany was a haven for dogs of various breeds, part of the German culture. For instance, there were short-haired dachshunds, long-haired, and wire-haired dachshunds. My wife, Nancy, and Jill Koele, the wife of another new officer, 2LT Rich Koele, both purchased dachshunds. Nancy chose a long-haired and Jill, a wire-haired, as shown in Figure 110A. In Figure 110B is our long-haired dachshund which we named Strudel. In Figure 110C is our family, Nancy, Strudel, and me, by our car outside of our Warner Kaserne Housing. Strudel was the first of four long-haired dachshunds we had over our time in the Army and beyond..Shown in Figure 111A is our battalion's new officers at the GTC at the annual Field Test Exercise (FTX) toasting with German beers in their quarters in Vilseck. Vilseck, Germany today is shown in Figure 111B.

It is interesting to note that our convoy from Munich to Vilseck in January 1968 was a lesson learned in driving on autobahns. In inclement weather conditions, the Autobahn E45/9, Figure 108A, was quite an experience. The autobahns had no speed limits, and the German cars would be traveling over 100 mph, often tailgating one behind the other. If one went off the highway, several followed. If we had a breakdown of one of our vehicles, we had to place a warning sign at least a mile down the Autobahn to get the German driver's attention.

Our FTX was successful. We were glad to return to Munich after the FTX. Two key veteran B Battery officers that I became close to early in 1968 were 1LT Daniel O' Brien, Executive Officer, and CPT Frank Gordon, CO. 1LT O'Brien was quite a character who always believed in having a good time. CPT Gordon was one of the few captains in the battalion, as most were in Vietnam. He and his wife, Mitzi, a native German, became close friends of Nancy and me, with whom we kept

Fig.110A. Different Dachshunds   *Fig. 110B. Strudel.*   *Fig.110C. Schrage Family*

in touch throughout most of our military and follow-on careers.

CPT Gordon helped me transition to CO, B Battery in 1968. One of the first things I had to do to become a Battery Commander was to be cleared for a Nuclear Duty Position. My screening by senior officers and

Figure 111A. New Officers Toast Beer in Vilseck

Figure 111B. Vilseck, Germany Today

supervisors found on 9 March 1968 that "The Officer is highly motivated and has the capabilities for such an assignment."

This was followed by my attending the one-week Honest John Warhead Pre-Fire Course. This was conducted at a Hawkins Kaserne, just outside of the small Bavarian village of Oberammergau, a municipality in the district of Garmisch-Partenkirchen in Bavaria, Germany. The small historic town on the Amer River, shown in Figure 112A, is known for its woodcarvers and woodcarvings, for its NATO School, and worldwide for its 380-year tradition of mounting Passion Plays every ten years. The next play is scheduled for Summer 2022. Nancy and I stayed in the Hotel Alte Post, shown in Figure 112B.

Figure 112A. Bavarian Village of Oberammergau.

Figure 112 B. Hotel Alte Post

After World War II, the Americans occupied the Hotzendorf Kaserne outside Oberammergau, renaming it Hawkins Barracks and making it the primary U.S. Army School Europe facility. Over the next three decades, schools in specialties ranging from Military Police to nuclear weapons handling were located there. However, the base reverted to German Army control and its original name in 1974. NATO School, formerly NATO Weapons Systems School, the alliance's principal training and education facility on the operational level, had been located at Hawkins Barracks/Hotzendorf Kaserne since 1953. The 1950s to the 1970s represented continuous change as U.S. forces across Europe consolidated their positions and rationalized facilities. This culminated in command of U.S. Army Schools being relocated to Oberammergau in 1960, remaining there until 1974. Shown in Figure 113A is the entrance into Hawkins Barracks, and Figure 113B shows the Barracks with mountains in the background.

Fig. 113A. Entrance to Hawkins Barracks.

Fig.113B. Hawkins Barracks w/ Mountains Background

WEAPONS ASSEMBLY SCHOOL

A vital element of the nuclear deterrent was to have a force at near-instant readiness to retaliate if attacked by the Soviet Union. U.S. forces had various tactical, and theatre nuclear weapons available, and constant training was required to ensure that they were stored and handled safely yet were available for deployment if tensions rose.

One weapon was the MGR-1B (M50) "Honest John" tactical nuclear missile, deployed across Europe from 1955 to 1975. At Oberammergau, missile operators were trained to prepare the warheads (principally installing the "physics package," fitting this to the launch body and the pre-fire sequence). Only "drill rounds" were used at Oberammergau – no nuclear warheads or chemical agents were stored here.

The Weapons Assembly Department moved from Piramens, south of Ramstein, to Oberammergau on 7 September 1960. On 1 October 1960, the school was re-designated as the Weapons Assembly Department of the U.S. Army Intelligence, Military Police, and Special Weapons School, Europe. It was further re-designated as the Weapons Assembly Department, U.S. Army School, Europe, on 1 July 1961, to reflect the overall re-organizing and concentration of the School at Oberammergau.

The mission of the Weapons Assembly Department was: *"to teach forward assembly, handling, and pre-fire procedures to personnel of delivery and support units within USA EUR and to familiarize commanders and selected staff personnel concerning required procedures and standards."*

The Department had five academic branches covering:

1. MIM-14 "Nike-Hercules" surface-to-air nuclear-armed missile
2. Atomic Demolition Munitions (Engineer)' nuclear land mines
3. MGR-1B (M50) "Honest John" tactical surface-to-surface nuclear missiles
4. W-48 & W-74 6 inch (155 mm)
5. M-33 8 inch (203 mm) nuclear artillery shells

I took the one-week Honest John Warhead Pre-Fire Course under the MGR-1B (M50) "Honest John" tactical surface-to-surface nuclear missiles and graduated as a "Superior Student," 4th out of 18 students. I understood the importance of this Course for my time as the CO for B Battery, 1st Battalion of 34th Artillery, and for my career, as will be discussed in my Stories. The Weapon Assembly faculty and staff are

Figure 114A. Weapon Assembly Faculty and Staff.   *Figure 114b.HJ Missile Loaded Demo*

shown in Figure 114A. Figure 114B shows an "Honest John" missile being placed on a launcher on the main parade ground & car park for an Open Day at Hawkins Barracks, Oberammergau, 1957.

While Munich, Germany, was an excellent assignment, some circumstances made me realize the challenges and importance of this assignment. They are summarized as follows:

1. Planning for Return of Forces to Germany (REFORGER)

- On Dec 20, 1967, HQ USAREUR (U.S. Army Europe) announced that most of the 24th Inf Division (division headquarters; two infantry brigades; division artillery; division support command), as well as several other Army and Air Force units, were to be redeployed to Fort Riley, Kansas. USAREUR also announced that HQ 3rd (Rotation) Brigade of the 24th Inf Division and 1st Bn, 34th Arty would be relocated to Augsburg starting in May and June 1968. The moves were part of a relocation plan connected with the more extensive REFORGER program (redeployment of the 24th Inf Division to the States).

- Exercise Campaign REFORGER was to become an annual exercise and campaign conducted by NATO during the Cold War. The exercise was intended to ensure that NATO could quickly deploy forces to West Germany in the event of a conflict with the Warsaw Pact. Most of the Army equipment from the units in Munich rotating to Fort Riley, KS, had to be prepared for storage in depot installations along the Rhein River.

- The maintenance units in Dachau and units moving to Augsburg were tasked to help prepare the vehicles for storage. However, in some cases, vehicle parts, bearings, and supplies were not available or were in short supply. Missing bearings were often packed with mud. Many of the officers and enlisted personnel did not realize that these vehicles would have to be pulled out of storage by the two brigades (1st and 2nd Infantry Brigades) 24th Infantry Division for participation in REFORGER I in less than a year. It will be described how many of the vehicles broke down on the autobahns when moving from the Rhein depots to the Grafenwöhr and Hohenfels training centers, GTC, and HTC.

2. Promotion to 1LTs and Battery Commanders

• On June 7, 1968, one year from their graduation date, the three USMA officers in the 1ˢᵗBattalion of 34ᵗʰ Artillery were promoted to 1LT and began training to become the Battery COs for A, B, and Headquarters Batteries. Battery COs are usually Captains. However, with the Vietnam War ongoing, few Captains were being assigned to U.S. Army Europe (USAEUR). 1LT Jan Askman became the A Battery CO; I became the B Battery CO, and 1LT Rick Hill became the Headquarters (HQs) Battery CO.

3. Czechoslovakia Crisis

• On the night of August 20, 1968, approximately 200,000 Warsaw Pact troops and 5,000 tanks invaded Czechoslovakia to crush the "Prague Spring"—a brief period of liberalization in the communist country. Although the Soviet Union's action successfully halted the pace of reform in Czechoslovakia, it had unintended consequences for the unity of the communist bloc. The invasion of Czechoslovakia by the Soviet Union in 1968 was one of the most extensive military operations on European soil since World War II. The attack was followed by a wave of emigration, including an estimated 70,000 Czechs and Slovaks initially fleeing, the total eventually reaching 300,000. The invasion sparked intense protests from Yugoslavia, Romania, China, and Western European communist parties.

REFORGER

I was initially scheduled for later in 1969. As a result of the Soviet invasion of Czechoslovakia in August 1968, the date was moved forward to reassure the NATO Allies that the U.S. could quickly reinforce Europe in a crisis. Because of stormy weather, seven means of transport were forced to put down at other bases short of their Rhein-Main destination. But dozens of others got through, delivering 447 tons of equipment and 2,058 troopers in three days.

Though the Soviets sent 200,000 soldiers into Czechoslovakia only five months before REFORGER I, they professed outrage at the comparatively modest influx of 12,000 U.S. troopers. Tass, the Soviet news agency, attacked REFORGER I as "a new provocative plot." Elaborating on that theme, Izvestia, Moscow's evening newspaper, warned that "the new military demonstration is directed at increasing tension in Europe." What bothered the Soviets most was that the war game would be held in Bavaria at the NATO maneuver site of Grafenwöhr —located only 30 miles from the Czechoslovak border.

MY STORY

Having completed the Honest John Warhead Pre-Fire Course, I received my Top Secret Security Clearance in June 1968 and was expected to take a more significant role in nuclear weapon operations. One of the most challenging tasks was as the Convoy Commander to move nuclear weapons from the NATO Nuclear Weapons Storage Site just north of Munich, Figure 105, to a new location.

Another officer was assigned as the Nuclear Weapons Officer for this relocation up to the new NATO Nuclear Weapons Storage Site near Ansbach, Germany. We did a very detailed route planning and visual assessment of the route we would take. Ground transportation of nuclear weapons in NATO was a great responsibility and required very high security, including an armed security ground force accompanying the movement. If the movement came to a halt, the security ground force was deployed on a perimeter around the convoy. Also, airpower was launched from NATO to help secure the nuclear weapons and protect the ground force. Our route required going through several small German villages with narrow village streets and small tunnels, typical of those shown in Figure 115.

Figure 115. Typical German Village Street and Tunnel

On our recon of the route, we found one tunnel we couldn't avoid, measured it in great detail, and thought we had an inch or two to spare. However, when we got there with the convoy, we found that it was an inch or two too small for one vehicle. Since we couldn't stop, we let the air out of the tires and went through the tunnel with flat tires. We were thankful when we made the rest of the trip unscathed. Later on, during REFORGER I in early 1969, the M88 Tank Retriever got stuck in a small village and backed up two battalions behind it. My B Battery was one of the units that were halted. I became more aware that the third dimension of air mobility was required.

MY STORY

By late Summer 1968, the remaining units of the 24th Infantry Division in Munich were moved to Augsburg, Germany. In many ways, it was good to move out of Munich. First, it was the spy capital of Germany. One of our Battalion officers was being court-martialed for living with a Russian Spy. Second, it was also the drug capital, as large drug supplies came to Munich from the Middle East. One night when I was the duty officer for the battalion, I was called to go down to the Munich Police Station to pick up one of our soldiers. I'll never forget the soldier was almost purple from such a large overdose, but he did survive.

RELOCATION TO AUGSBURG, GERMANY

Figure 116A. Sheridan Kaserne Augsburg Gate.   *Figure 116B. Buildings in Dawson Field Area.*

Fig. 117A. 1st Bn 34th Arty.

Our movement to Augsburg was timely and welcomed. One reason for the movement out of Munich was that Germany was beginning to prepare for the 1972 Munich Olympics, and the Germans didn't want to look like an occupied country. The 24th Infantry Division Munich elements moved into the Sheridan Kaserne in Augsburg. Shown in Figure 116A is the Entry Gate to Sheridan Kaserne. Shown in Figure 116B are buildings in the Dawson Field Lawn Area, making it one of the most beautiful kasernes in Germany.

Due to its overall appearance, Sheridan Kaserne was always unique with regard to the other kasernes. Last but not least, cultural as well as military events took place in the Officer's Club, where, at certain times, important decisions were made. The 600 meters long and 80 meters wide Dawson Field lawn area between buildings (a former mess hall, then Child Care Center and Bavaria House mess hall) in the south of the Kaserne allowed for grand parades and helicopter training during the Cold War confrontation. It also was used as an athletic field, including a tennis court. There was no comparable area in Augsburg that offered such an uninhibited view. The 1st Bn 34th Artillery location in Sheridan Kaserne with my B Battery practicing assembly of an HJ Rocket on the launcher is shown in Figures 117A, B, and C with 1LT Schrage identified in Figure 117B.

Figure 117B. ILT overseeing Assembly   *Fig 117C. Final Assembly*

Since the three 1LT USMA graduates were now Battery Commanders, the Battalion Commander LTC Pheifer put us in new quarters adjacent to the Sheridan Kaserne. These quarters were close to him and his battalion staff, as numerous challenges were forthcoming, and the required joint teamwork between the three Batteries and the Battalion Staff was essential. Shown in Figure 118 are examples of the new officer quarters for the Battery Commanders and Battalion Staff.

Figure 118. Example Officer Quarters in Sheridan Kaserne

With the Soviet Union in Czechoslovakia in late summer and early fall, there were numerous more nuclear weapons Technical Proficiency Inspections (TPIs) from three higher headquarters, as illustrated in Figure 119. USAREUR/Seventh Army, VII Corps, and the 24th Infantry Division, our parent organization. The TPIs were conducted to ensure that units with nuclear weapons were ready, especially since the Soviets and their tanks were fewer than several hours away.

A nuclear weapons unit had fifteen minutes to open the safe and decipher classified messages to see if they were real or false alarms. It took two teams, a Red Team and Blue Team, to open the safe and decipher the messages. These teams usually had two or more team members.

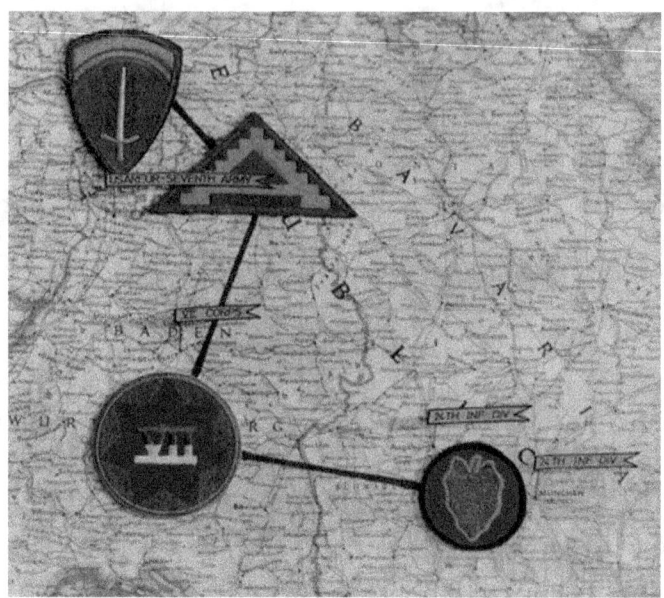

Figure 119. Higher Headquarters Required Numerous TPIs for Nuclear Weapon

With the rapid turnover of officers in Vietnam, there were usually only two members per team. All had to have Top Secret clearances with special access that took several months to obtain.

MY STORY

There was a period for a few months in early Fall 1968 that I was the only eligible member of the Blue Team. It meant that I could never be more than fifteen minutes away from opening the safe and deciphering the message in Battalion Headquarters. My parents had planned to

Figure 120A. Nancy and Dan's parents in Italy.

Figure 120B. Marnen by River

make a trip and visit us in the Summer of 1968 when they would be out of school, as they were both still teaching. It was before I became the only Blue Team member in Fall 1968. Therefore, Nancy and I were able to make a quick trip through Switzerland and Italy with my parents.

Shown in Figure 120A is a picture of the leaning tower of Pisa, Italy, with Nancy and my parents standing next to it while I took the picture. When Nancy's parents came to visit in the Fall of 1968, I was the only Blue Team member. Therefore, Nancy took them on a trip to Germany and Switzerland. Her father, Marnen, wanted to visit where he was wounded in World War II. His unit was crossing a river in Germany. Figure 120B shows Marnen standing next to the river where he was injured and lost his leg to a booby trap.

One unique event during my period as the only Blue Team member was that I could access the golf course, which adjoined Sheridan Kaserne. Today, it is shown relative to the Sheridan Kaserne Housing Area, where we lived in Figure 121A. I had a jeep on standby at the golf course and got called off the golf course twice to decipher the classified messages in less than fifteen minutes. The golf course was a German golf course, but they allowed U.S. service members to play on it. I remember playing in a joint U.S.- German tournament on it. The Germans were good golfers, but they drank a lot of beer as they played, and usually, by the back nine of the golf round, their shots weren't as good, and we could catch up with them.

The Sheridan Kaserne was named after PFC Carl V. Sheridan, who had been posthumously awarded the Medal of Honor in World War during the attack on Frenzenberg Castle (now Burg Frenz today). The Sheridan Kaserne is shown in Figure 121B.

Figure 121 A. Golf Course by Sheridan Kaserne.

Figure 121B. Burg Frenz today

RACIAL UNREST AND VIETNAM PROTESTS

In 1968-69, there was substantial unrest in the USA regarding racial unrest and anti-Vietnam War protests and demonstrations. I remember having to help break up a confrontation between blacks and whites in Sheridan Kaserne one night in the Fall of 1968. While the Vietnam War protest wasn't as evident in Germany except on television, the troop buildup in South Vietnam impacted the lack of experienced officers, especially at the rank of Captain in Germany.

REFORGER 1

REFORGER was not merely a show of force—in the event of a conflict, it would be the actual plan to strengthen the NATO presence in Europe. In that instance, it would have been referred to as Operation REFORGER.

Exercise Campaign REFORGER became an annual exercise and campaign conducted by NATO during the Cold War. The training was intended to ensure that NATO could quickly deploy forces to West Germany in the event of a conflict with the Warsaw Pact. REFORGER I was called *Carbide Ice,* and the official Exercise was from January 29 – to February 4, 1969. The advance party for the training exercise REFORGER I arrived at Rhein-Main Air Base in Germany on January 6, 1969, Figure 122A. The REFORGER I Operation included the redeployed units, such as the 1st and 2nd Infantry Brigades from the 24th Infantry Division, Fort Carson, CO, as a Blue Force based out of the GTC. The 3rd Infantry Brigade of the 24th Infantry Division and the 1st Battalion 34th Artillery Division Artillery were from the Sheridan Kaserne and part of the Red Team based out of the Hohenfels Training Center (HTC). The Operation Area (Mock War) was conducted over the German countryside between the two training centers, as illustrated in Figure 122B. It was determined that REFORGER I would succeed, although there were several detriments to achieving it.

Figure 122A. Advance Party for REFORGER I.

MY STORIES

While REFORGER I was declared a success, several circumstances hindered this conclusion. First, many of the deploying units stored equipment in the Rhein depots never made it to the Grafenwöhr Training Center (GTC) without breaking down along the Autobahn. As I recall, one 8-inch artillery battery only made it to GTC with two of its eight howitzers.

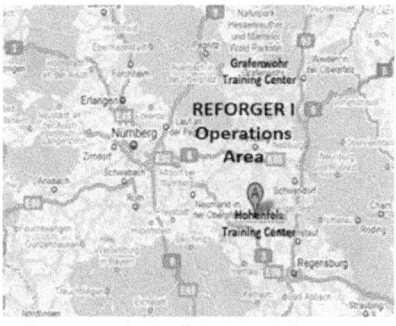

Figure 122B. REFORGER I Operations Area

Another problem was that the weather was exceptionally mild, above freezing, for that time of the year. It caused two detriments. One was units, such as my HJ B Battery, got stuck in the mud. Our large transport trucks, which carried our HJ missiles/rockets, were gasoline-powered with little power for traction in the mud.

Another problem was that while we had two wreckers, they could only be used as special weapons wreckers to mate the warheads to motors on the launchers, as in Figure 123A.

Figure 123A. Special Weapons Wrecker Mounting Warhead.

Figure 123B. HJ Setup in the Field

Figure 124. 1LT Schrage and SFC Bohannon having a beer in a German Gasthaus

Being division artillery, my B Battery could often choose its relocation site independent from the Battalion Headquarters, as shown in Figure 123B. We had to relocate to a new position almost daily. Our troop's morale was extremely low, having to dig out our vehicles daily. To remedy this, my First Sargent, SFC Bohannon, and I came up with a remedy for this situation. As we passed through some small German villages, we noticed that the German beer houses, i.e., Gasthausen, were advertising Fasching parties. Fasching is Germany's carnival season, similar to Mardi Gras. It begins on the 11th day of November at precisely eleven minutes after 11 am and ends at the stroke of midnight on Shroud Tuesday. It is often called Fat Tuesday (the Tuesday before Ash Wednesday). This is the timeframe of REFORGER I. Fasching is more or less a Roman Catholic and Christian Orthodox celebration, and most Protestant and non-Christian areas do not celebrate it. Therefore, as we looked for a new site to set up our Battery, we would locate it close to a German village advertising a Fasching Party.

Shown in Figure 124 is a picture of SFC Bohannon and me sitting in a German Gasthaus during REFORGER I.

If we stayed in the exact location for two nights, SGT Bohannon would take half of the Battery to the Fasching Party the first night, and I would take the other half the second night. We had to set up some security operations to keep from being picked up by the Military Police (MP), who were patrolling much of the REFORGER I Operations Area. We did this by putting a hidden soldier at each end of the village with a PRC 25 Radio to warn us if an MP jeep was approaching the village. Fortunately, no encounters took place. The troops' morale greatly increased and made the rest of the exercise enjoyable.

One other event was significant to me regarding LTG Donald Bennett. The VII Corps Commander was overall in charge of REFORGER I. He had his staff contact our Battalion Headquarters, as he wanted to visit my B Battery site in the field. He had been the USMA Superintendent who handed me my diploma, Figure 87B, Chapter 3. I had to make sure my Battery was ready for his visit. It went well, and he was complimentary about our layout and my description of our operations. As he was leaving, he sent his aide, a Captain, to inquire if I was interested in being LTG Bennett's aide. I told him to thank LTG Bennett, but I wanted to serve out my time as Battery Commander. Also, I would be going to rotary-wing flight school later in Spring 1969. My desire to go to flight school was reinforced by my REFORGER I experiences. It was evident that air mobility was required on the future battlefield. Due to the weather conditions, the armor units in REFORGER I were often restricted to hard surface roads. Figure 125A. When they tried to go through the small German villages, the tanks and tank retrieval vehicles often got blocked by several battalions lined up one behind the other, Figure 125B.

Figure 125A. Armored Vehicles Restricted to Roads.

One of the big complaints from the German residents about REFORGER I was the damage to their countryside. It was estimated that there were more than a million dollars a day in maneuver control damage to the German countryside, including the forests, which in Germany are like parks.

Figure 125B. Narrow Streets Blocked Tanks

My experiences in REFORGER I were outstanding. Being able to lead and position an Honest John Rocket/Missile Battery in a battlefield environment was a great lesson learned which would serve me well in Vietnam. Learning about nuclear weapons and what they entail was also very valuable.

After REFORGER, my return to Augsburg gave me time to spend with my wife, Nancy, who was pregnant with our first child. There was one more field exercise back to GTC in March 1969 for the 1st Battalion 34th Artillery annual FTX, which was moved from January to March due to REFORGER I. While my Battery performed well, we did not put the HJ rocket on the required football field due to a crew member's error in setting the required elevation. This would follow with my return to the USA for Rotary Wing Flight School once I received orders. Since Nancy wouldn't be able to travel six weeks before our baby was due, she flew back to her home in Carlyle, IL, in late February 1969. Our first child, Steven, was born on March 31, 1969. Fortunately, I received my orders to flight school in early March 1969; however, I couldn't fly back to the USA until the first week of April 1969. Therefore, I missed the birth of my son, Steven, who was born at Scott AFB Hospital in Shiloh Valley, Illinois. My absence would occur again with the birth of our daughter, Susan, born in November 1970, while I was serving in the Vietnam War.

Rotary Wing Flight School & Deployment to Vietnam

"Hey it's good to be back home again, yes it is"- John Denver

It was great to return home from Germany and see my newborn son and visit with the rest of our family. We were then off to Fort Wolters, TX, to begin the four-month primary phase of rotary-wing primary helicopter training. This would be followed by the four-month advance phase at Fort Rucker, AL, or Fort Hunter-Stewart Army Airfield, near Savannah, GA. Deployment of Vietnam would follow.

Primary Helicopter Flight Training at Fort Wolters, TX

Fort Wolters, TX, was located next to Mineral Wells, TX, a little more than one hour west of Fort Worth, TX, as illustrated in Figure 126A. The entrance to Fort Wolters is in Figure 126B.

Mineral Wells, TX is named for mineral wells in the area, which were highly popular in the early 1900s. In 1919, Mineral Wells hosted the spring training camp for the Chicago White Sox, the year of the famous "Black Sox" scandal involving "Shoeless" Joe Jackson. Mineral Wells also hosted spring training for the Cincinnati Reds and St. Louis Cardinals in the 1910s and early 1920s. The baseball field was in the center of town, where a shopping center now sits.

In 1956, Camp Wolters reverted to the United States Army to house the United States Army Primary Helicopter School (USAPHS). In 1963 it was designated a permanent military base and renamed Fort Wolters.

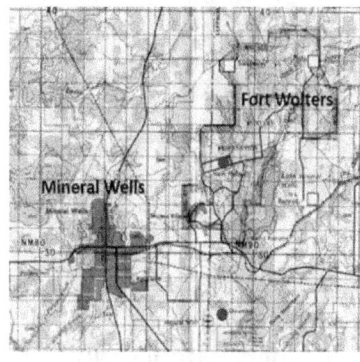

Figure 126A. Mineral Wells and Fort Wolters Map.

Figure 126B. Fort Wolters Entrance

The facility started with one heliport (Main) and 4 stage fields. At its height, it had three heliports (Main Heliport, Downing Field, and Dempsey Field) and twenty-five stage fields (Pinto, Sundance, Ramrod, Mustang, Rawhide, Bronco, Wrangler, An Khe, Bac Lieu, Ben Cat, Ben Hoa, Cam Ranh, Can Tho, Chu Lai, Da Nang, Hue, My Tho, Phu Loi, Pleiku, Qui Nhon, Soc Trang, Tay Ninh, Tuy Hoa, Vinh Long, and Vung Tau). The Vietnamese-named stage fields were named after facilities in Vietnam and were oriented to be the same relation to each other, on a smaller scale, of course, as they were on the map. The other stage fields were Western-themed. In June 1963, the post was re-designated Fort Wolters, a permanent military installation and U.S. Army Primary Helicopter Center.

When the school at Fort Wolters was in operation, it shared the total effort for training helicopter pilots to fulfill the Army's requirements. This effort was shared with the Army Aviation Schools at Fort Rucker, Alabama, and the Hunter Army Airfield, at Fort Stewart, Georgia.

All potential Army helicopter pilots received their primary training at the USAPHS at Fort Wolters. First, they trained for sixteen weeks (20 weeks for Warrant Officer Candidate students who underwent a four-week WOC indoctrination course prior to flight training).

Afterward, they went to Fort Rucker for advanced helicopter training for 16 weeks. Courses offered at the school included the following two courses:

- The 16-week Officer / Warrant Officer Rotary Wing Aviator Course (ORWAC) was designed to train officers with no previous military flying experience.
- The Warrant Officer Candidate Rotary Wing Aviator Course (WORWAC) (the first four weeks were the WOC Indoctrination Training Course - commonly known as "Pre-Flight").

The first eight weeks of the students' flight training were devoted to learning the basic flight maneuvers of the helicopter at a fixed operation area on a "stage field." This training was conducted under civilian contract by Southern Airways of Texas, Inc., and Flight Department "C" of USAPHS.

The second eight weeks, or Primary II, taught students to apply the basic flight maneuvers to small unimproved landing areas and introduced formation flying, air navigation, and night flying. This training was conducted by Flight departments "A" and "B" of USAPHS. The military instructors of these divisions were nearly all combat veterans.

TH-55A Helicopter in Armed Forces Day Activities

Figure 127B. TH-55A Osage

Figure 127A. OH-23D Raven.

TH-55A Helicopter in Armed Forces Day Activities

Figure 127C. OH-13 Sioux

The School started with 125 Hiller OH-23D Raven helicopters (Figure 127A). The number peaked in 1969 at more than 1300, including the OH-13 (Figure 127B) and TH-55 types (Figures 127C and D). Over 41,000 students, representing over 30 countries, graduated from the primary helicopter school during the 17 years it functioned in this capacity.

Peak output occurred in 1967, with 600 students graduating each month. Put in "mothballs" in November 1973, what was once known as Fort Walters is now Wolters Industrial Park, housing a number of businesses, and is part of the old Warrant Officer Candidate (WOC) area.

The TH-55A Osage became the primary training helicopter, Figure 127C and D. Larger persons such as me trained in the OH-23D, Figure 127A.

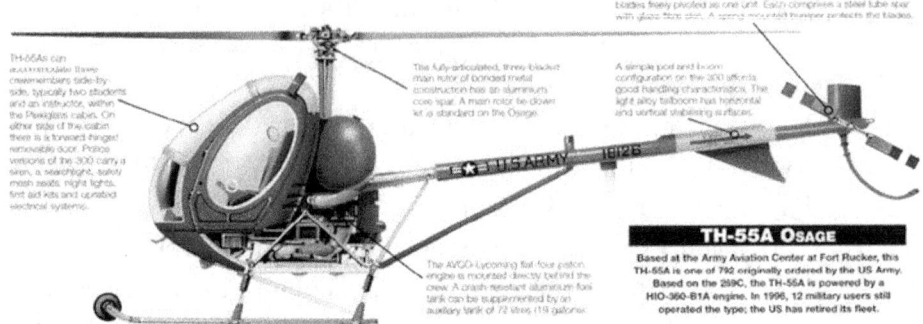

Figure 127D. The Army Primary Training Helicopter, The TH-55A Osage

MY STORIES

My family, Nancy, Steven, and I, arrived in Mineral Wells, TX, and Fort Walters in late April 1969. We rented a house in Mineral Wells, TX.

Also, I began learning to become a helicopter pilot. It was not easy for me. The first step was ground school instruction. Following ground school instruction, one had up to ten hours of flying with an instructor pilot before flying solo. If you didn't, you were either sent back to the beginning of the dual flying with the instructor or flunked out of flight school. I got to nine hours of dual flying with the instructor and wasn't sure I would be able to do solo. However, the instructor pilot came out for my tenth hour of flying and told me it was time to solo, which I did successfully. I found that the skills required to fly a helicopter were not athletic but more instinctual, like learning to play the piano. I got my wings put on by two of my classmates, as illustrated in Figure 128A.

Figure 128A, Classmates giving me my Wings

166

Figures 128B, Class 69-44 winning Softball Team

These two classmates were also part of our USAPHS 69-44 Class Fast Pitch Softball Team, which won the Fort Wolters Post Championship Team. The Team is shown in Figure 128B. It was the first time that a student team had won the Post Championship, as student teams rolled over during the season. I am on the left end of the back row and was a starting outfielder and backup pitcher.

Figure 129, USAPHS Class of 69-44 Graduating Class

I was promoted to Captain on June 7, 1969, the third anniversary of my graduation from USMA. The USAPHS 69-44 Class is shown in Figure 129.

When Nancy and I went to pay our water bill when checking out our rental house in Mineral Wells, TX, we were in for a shock as it was for several hundred dollars. While it had been an unusually wet August month for Mineral Wells, TX, and some water was sitting in our backyard, we didn't realize that we had a broken water pipe. Fortunately, the owner of the house paid this unrealistic water bill.

U.S. ARMY ADVANCE COURSE HELICOPTER TRAINING AT FORT RUCKER, AL

Graduates from the USAPHS chose to take advanced helicopter training at either Fort Rucker, AL, or Hunter Army Airfield at Fort Stewart, GA, near Savannah, GA. Nancy and I opted for Fort Rucker, AL. It was a shame the class was broken up as many of our close friends, especially the bachelors on the softball team, opted for Hunter Army Airfield at Fort Stewart, GA. The helicopter Advance Course training was less intensive than the Primary Training, as we now knew how to fly. It concentrated more on instrument flight and tactics training. We also realized that what we were learning would soon be real in Vietnam after graduation.

The Rotary Wing training at Fort Rucker began in the Department of Rotary Wing Training. The first four weeks were in the Basic Instrument Division at Shell AAF (designed initially as a Fixed Wing Field). The student went to the Advanced Instrument Training Division at Hanchey AHP. At Hanchey, the TH-13T (Figure 130A), and

Figure 130A. TH-13T IFR
Training Helicopter.

Figure 130B. TH-13T Instrument Panel

instrument panel (Figure 130B), was the helicopter's instrument trainer and, later, the Huey (UH-1). Initially, the training objective was to provide enough instrument experience to get out of trouble if a pilot encountered "inadvertent IFR." However, as soon as possible after the Vietnam War, the Advance Course extended training to provide a full-up Instrument qualification and issued a "Standard Instrument Card." This qualification was called a "Tactical Instrument Card."

Students spent many hours in the "Link" trainer during this phase and spent much time under the hood. The "hood" was a device placed over the top of the eyes, like a blinder on a horse, to prevent any outside vision other than the instruments.

The Next Phase was with the Contact Training Division at Knox AHP. Here students learned how to take off and land safely, carry heavy loads, and do autorotations (power off) landings. That provided a four-week transition into the Huey.

For the first two weeks, students staged to Tac-1 and moved to Tac-x for the final two weeks. The last Phase was with the Department of Tactics (which later became the Tactics Training Division of the Department of Rotary Wing Training). That training operated out of Lowe AAF. The tactical training phase attempted to provide Vietnam-oriented experience as closely as possible. Until 1972, at least every instructor pilot assigned to "Tactics" was a combat returnee from Vietnam. They knew firsthand what the new Aviators would soon face, and they gave it all they had! There was never a more devoted group of instructor pilots than those assigned there.

MY STORIES

Fort Rucker is a U.S. Army post located primarily in Dale County, Alabama, United States. It was named for a Civil War officer, Confederate General Edmund Rucker. The Post is now the primary flight training installation for U.S. Army Aviators. It is home to the United States Army Aviation Center of Excellence and the United States Army Aviation Museum. Small sections of the Post also lie in Coffee, Geneva, and Houston counties. Part of the Dale County section of the base is a census-designated place; its population was 4,636 at the 2010 census.

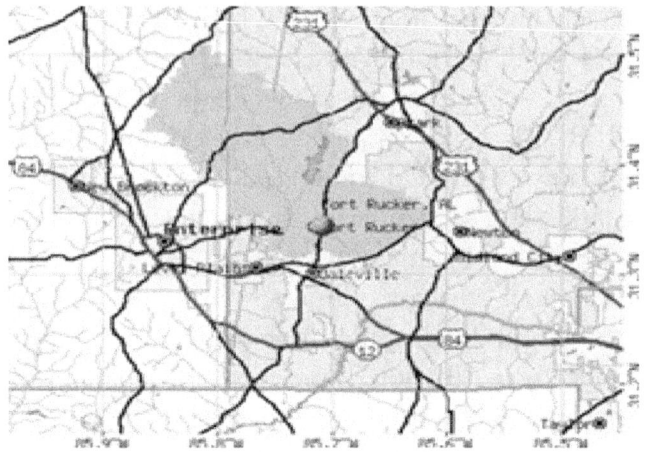

Figure 131A. Map of Fort Rucker & Surrounding Area.

A map illustrating the Fort Rucker training center and surrounding communities are shown in Figure 131A. The Fort Rucker Entrance Gate is shown in Figure 131B.

131B. Fort Rucker Entrance

Nancy and I decided to rent a house trailer in Daleville, AL, due to its lower cost and proximity to Fort Rucker.

Today, Daleville is a city in Dale County, United States. At the https://en.wikipedia.org/wiki/2010_United_States_Census"At" https://en.wikipedia.org / wiki / 2010 United States Census" At the 2010 census, the population was 5,295, up from 4,653 in 2000. It is part of the "https://en.wikipedia.org/wiki/Dale County, Alabama" Ozark Micropolitan Statistical Area.

Daleville's nickname is "Gateway to Fort Rucker," as this U.S. Army post is just north of town. Cairns Army Airfield, the major Fort Rucker Airfield, is located to the south of Daleville on the road to nearby Clayhatchee.

Figure 132. Dan, Nancy & Dan's Parents, Albin and Mary, at St Louis West Point Parents Dinner

We were fortunate to have Christmas Week off before returning to Fort Rucker to complete our Advanced Course and Graduation. We returned to Southern Illinois and were fortunate to attend a formal dinner and dance sponsored by the St. Louis West Point Parents Club with my parents, as shown in Figure 132.

When we returned to Daleville, AL, and Fort Rucker, we prepared for Vietnam deployment within two weeks of graduation. We had made many friends in Flight School. Most were also headed for Vietnam in January 1970. One of the major social events at the Fort Rucker Officer's Club was a weekly Bingo in which prizes for filling full cards were boats, cars, and airplanes. At the last Bingo Nancy and I attended, I was the co-full card winner for a new Opel Kadette. We were given the choice of splitting the prize or drawing cards, with the highest card winning the car. I decided to draw a card and won the car. Fortunately, we could drive the vehicle to Nancy's parents' home in Illinois, and they were able to sell it.

Graduation from flight school consisted of a graduation ceremony, as illustrated in Figure 133A. Also, there was a formal ball at the Officer's Club. Shown in Figure 133B is a picture of Nancy and me at the graduation ball.

Figure 133A. Flight School Graduation   *Figure 133B. Flight School Graduation Ball*

Deployment to Vietnam

Following graduation, Nancy and I packed our car and drove to her parent's home in Carlyle, IL. We had rented a house for Nancy and our son, Steven, only a few blocks from her parents' home, while I was in Vietnam. My parents lived about twenty minutes away in St. Rose, IL.

I flew from St. Louis, MO, to Oakland, CA, to catch a chartered aircraft flight to Tan Son Knut Air Base in Saigon, South Vietnam, for deployment to Vietnam. Two close friends from Flight School joined me in Oakland, CA, the day before the chartered flight left for South Vietnam.

One of them was Terry Mote. He and his wife had become close friends with Nancy and me. Unfortunately, Terry was the first flight school classmate killed in Vietnam when he lost an engine during take off and crashed. Many of us were aware of the danger of flying helicopters in the Vietnam War. However, we felt we were essential in saving lives and winning the War. Approximately 40,000 U.S. Helicopter Pilots flew in the Vietnam War. About **2,202** pilots were killed, along with 2,704 crew members. For those with their hands on the controls, this meant 5.5% never made it back. Considering that the average pilot flew four times a week, he could expect to be flying up against the Grim Reaper on 11.4 of his flights during his tour in Vietnam. That means that every 4.5 weeks, he faced death. In soldier talk, his life expectancy was four and a half weeks; basically, a month.

SUMMARY OF BOOK 1: AS USUAL GUARDIAN WAS PERFECT IN ALL REPECTS

The objective of this book was to illustrate how the foundation of my first career as an active-duty Army officer was built on a strong family support element. It was also fostered by my athletic achievements and civil and military education opportunities. The tours in Germany as a Nuclear Weapons Battery Commander and in the Vietnam War as an assault helicopter pilot, platoon leader, air mission commander, and S-3 operations officer provided unique experiences. These assignments supplied leadership development and an understanding of operational opportunities.

They also raised the main issue from my first career on *Why Winning the Vietnam War was not successful.* As I suggested in Book 1 of my Trilogy, Lewis Sorley's book, *A Better War,* shows that from 1968 to 1972, a better war could have been extended to win the Vietnam War. I suggested that an opportunity came available to accomplish this goal when General Lon Nol took over the government in Cambodia to make it a democracy. To set the record straight, this was an opportunity that the U.S. Government turned down. This resulted in the U.S. not only losing the war in Vietnam but causing genocide in Cambodia. This was when the Pol Pot-led Khmer Rouge took over the country and deliberately killed millions of Cambodian people to destroy the country's intellectual, religious, and innocent people. These experiences would help me transition and advance in my subsequent careers in the military, civil service, and academia.

Book 2 in the Trilogy will begin with my return from Vietnam and assignment to the Field Artillery Officer Advance Course, followed by advanced civil schooling at Georgia Tech. This led to my second career as an aerospace engineer, technology manager, and senior executive directly involved in developing *the next generation of army aviation systems.*

AFTERWORD

A FULL LIFETIME CAREER OF SEEKING PERFECTION DRIVEN BY FAMILY AND MENTORS: A TRILOGY

Dr. Daniel P. Schrage
Trilogy Summary

The trilogy is a memoir of the author's three careers of military achievement, engineering accomplishment, and academic leadership. It documents how the author has strived for perfection through athletic and career advancement while presenting how he supports, follows, and documents the changes in U.S. warfare, technology development, and academic transition in a changing world. It will also try and set the record straight for these changes based on the author's three careers.

The trilogy begins with Book 1, *As Usual, Guardian was Perfect in All Respects,* with changes in the U.S. military as it transitions from the Cold War in Europe to the strategic mobility and nation building efforts of the Vietnam War.

It follows with Book 2, *Development of Next Generation of Army Aviation Systems,* as it documents and provides for the growth of air mobility through technology development over the next decade. It also attempts to set the record straight on why the development of Army Aviation systems has been so difficult.

It then transitions into Book 3. *Establishing a Graduate Program in Aerospace Systems Design.* It is based on focusing technology for Affordability through Integrated Product and Process Development (IPPD). In Book 3 these technological advancements are brought into an academic environment with innovative methods for education transfer to both the civilian and military students and industry. The use of IPPD for developing and demonstrating autonomous unmanned aerial vehicles (UAVs) is also presented. This setting also brings to life the changes in education at Georgia Tech with a graduate program in Aerospace Systems Design. Also, IPPD is used in Science Technology, Engineering, and Mathematics (STEM) training programs for high school students with NASA and industry support. It includes assistance for the United States Military Academy (USMA) in transitioning from a general engineering degree to the adoption of major engineering degrees, e.g., electrical, mechanical, and systems. Also provided is how the Georgia Tech graduate education and professional development programs help industry and government with the transition to Integrated Product and Process Development (IPPD) as they respond to worldwide competition and the Japanese Total Quality success.

These three books are connected by a "Why," which focuses on a strive for perfection. They also follow a growing Schrage Family with numerous relocations during the early years while striving for stability during later years. They will also try and set the record straight for each of the author's careers by answering the following questions:

First Career, Book #1: Why Winning the Vietnam War was not successful

Second Career, Book #2: Why is it so difficult for the Army to develop New Aircraft

Third Career, Book #3: Why is it so hard for Academia to develop new Curriculum

Book 1 in the trilogy is available to order at your favorite local bookstore or online wherever books are sold. Books 2 and 3 are scheduled to be released in late 2022 and 2023 respectively. You can go to the author's website at DanielSchrage.com for more information and release dates.

Acknowledgments

This is the first book of my memoir A Full Lifetime Career of Seeking Perfection Driven by Family and Mentors - A Trilogy

First of all, I would like to acknowledge my family for their support and dedication to me in helping document my full lifetime career in writing this trilogy. My wife, Nancy, has been with me since we started dating in high school in 1961 and has shared in the unique experiences and stories in the trilogy for the past 60 years. Chapter 3 in Book 1: Growing Up and Moving Forward clearly states her support and dedication as well as that of my entire family, father, mother, and sisters.

I would also like to acknowledge my friends and colleagues who have encouraged me to write this trilogy. In particular, Dr. Bill Lewis, former Army Aviator and Senior Executive Servant (SES), LTC (Ret) Paul Fardink, former Army Aviator and Historian, and Mr. George Singley, former Army SES and VP SAIC. They also served as proofreaders for some of the chapters.

I would also like to acknowledge the mentors who advised me in my three careers and gave me the resources and encouragement to succeed.

Finally, I would like to thank my publisher, Frank Eastland, and his team— particularly Raeghan Rebstock, Bob Laning, and Teresa Evans—at Publish Authority for all their help in bringing my memoir to reality.

About the Author

Dr. Daniel P. Schrage grew up in the Midwest as the son of two teachers and graduated from USMA West Point in 1967. He has advanced degrees in Aerospace Engineering (AE), MS Georgia Tech, 1974; Business Administration, Webster College. 1975; and a DSc. Mechanical Engineering (ME), Washington U. (St. Louis) in 1978.

Dr. Schrage has had three careers and excelled in each. In his first career, he commanded an Honest John Nuclear Missile Battery during the Cold War in Europe, 1968-69. He was an Army Aviation Air Mission Commander in South Vietnam and commanded lift ship and gunship platoons. He then served as the S-3 3th Combat Aviation Battalion and ran all operations in the Mekong Delta. He orchestrated the transfer of the first Army Aviation Airfield, Soc Trang, to the Vietnamese Air Force (VNAF) under the Helicopter Vietnamization Program, 1970-71. In his second career, he was an Aerospace Engineer, Manager, and Senior Executive for Army Aviation Development, 1974-78. He led the technical development of the next generation of Army Aviation Systems as the youngest Senior Executive Servant (SES) in the Army Material Command (AMC). In his third career, he was a Full Professor and Director of Research for the Army Vertical Lift Research Center of Excellence (VLRCOE) at Georgia Tech. He retired as a Professor Emeritus. His careers followed and documented the changes in U.S. warfare, technology development, and academic transition in a changing world. He is a Technical Fellow of both the American Institute of Aeronautics and Astronautics (AIAA) and the Vertical Flight Society. He and his wife reside in Atlanta, GA.

You can find more information about the author and his books on his website at DanielSchrage.com.

References for Book 1: As Usual, Guardian Was Perfect in All Respects

1. Song Titles and Some Lyrics in chapters from John Denver's Definitive All-Time Greatest Hits.

2. Vietnam Studies: Air Mobility, 1961-1971, LTG John J. Tolson

3. Sorley, Lewis., "A Better War: The Unexamined Victories and Final Tragedy of America's Last Years in Vietnam," 1999

4. Sorley, Lewis., "WESTMORELAND- The General Who Lost Vietnam," 2011

5. Foreign Relations, 1969-1976, Volume VII. Vietnam July 1970-January 1972. Special National Intelligence Estimate 1SNIF 57-70 was released in SEPTEMBER 2010.

6. A STORY ON SPECIAL FORCES OPERATIONS (NUI COTO AND THE MIKE FORCE ASSAULT ON TUK CHIP0, AN EXAMPLE OF SPECIAL FORCES OPERATIONS, EARLY 1969 (Wikipedia and Stormbringer, MONDAY, NOVEMBER 17, 2014)

7. A STORY ON AMERICAN POWS LEFT BEHIND (REPORTED U.S. PRISONERS UMINH FOREST). Joint Rescue Casualty Center (JRSC) Case #3115: "The Western Caucasian Prisoners Rumored to be American POWs Seen Being Escorted into the U Minh Forest (Palm Forest of Darkness) in Southern Vietnam in 1978."